Springer Tracts in Modern Physics
Volume 186

Managing Editor: G. Höhler, Karlsruhe

Editors: H. Fukuyama, Chiba
J. Kühn, Karlsruhe
Th. Müller, Karlsruhe
A. Ruckenstein, New Jersey
F. Steiner, Ulm
J. Trümper, Garching
P. Wölfle, Karlsruhe

Honorary Editor: E. A. Niekisch, Jülich

Now also Available Online

Starting with Volume 165, Springer Tracts in Modern Physics is part of the Springer LINK service. For all customers with standing orders for Springer Tracts in Modern Physics we offer the full text in electronic form via LINK free of charge. Please contact your librarian who can receive a password for free access to the full articles by registration at:

http://link.springer.de/series/stmp/reg_form.htm

If you do not have a standing order you can nevertheless browse through the table of contents of the volumes and the abstracts of each article at:

http://link.springer.de/series/stmp/

There you will also find more information about the series.

Springer
Berlin
Heidelberg
New York
Hong Kong
London
Milan
Paris
Tokyo

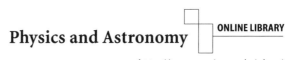

http://www.springer.de/phys/

Springer Tracts in Modern Physics

Springer Tracts in Modern Physics provides comprehensive and critical reviews of topics of current interest in physics. The following fields are emphasized: elementary particle physics, solid-state physics, complex systems, and fundamental astrophysics.
Suitable reviews of other fields can also be accepted. The editors encourage prospective authors to correspond with them in advance of submitting an article. For reviews of topics belonging to the above mentioned fields, they should address the responsible editor, otherwise the managing editor.
See also http://www.springer.de/phys/books/stmp.html

Managing Editor

Gerhard Höhler

Institut für Theoretische Teilchenphysik
Universität Karlsruhe
Postfach 69 80
76128 Karlsruhe, Germany
Phone: +49 (7 21) 6 08 33 75
Fax: +49 (7 21) 37 07 26
Email: gerhard.hoehler@physik.uni-karlsruhe.de
http://www-ttp.physik.uni-karlsruhe.de/

Elementary Particle Physics, Editors

Johann H. Kühn

Institut für Theoretische Teilchenphysik
Universität Karlsruhe
Postfach 69 80
76128 Karlsruhe, Germany
Phone: +49 (7 21) 6 08 33 72
Fax: +49 (7 21) 37 07 26
Email: johann.kuehn@physik.uni-karlsruhe.de
http://www-ttp.physik.uni-karlsruhe.de/~jk

Thomas Müller

Institut für Experimentelle Kernphysik
Fakultät für Physik
Universität Karlsruhe
Postfach 69 80
76128 Karlsruhe, Germany
Phone: +49 (7 21) 6 08 35 24
Fax: +49 (7 21) 6 07 26 21
Email: thomas.muller@physik.uni-karlsruhe.de
http://www-ekp.physik.uni-karlsruhe.de

Fundamental Astrophysics, Editor

Joachim Trümper

Max-Planck-Institut für Extraterrestrische Physik
Postfach 16 03
85740 Garching, Germany
Phone: +49 (89) 32 99 35 59
Fax: +49 (89) 32 99 35 69
Email: jtrumper@mpe-garching.mpg.de
http://www.mpe-garching.mpg.de/index.html

Solid-State Physics, Editors

Hidetoshi Fukuyama
Editor for The Pacific Rim

University of Tokyo
Institute for Solid State Physics
5-1-5 Kashiwanoha
Kashiwa-shi, Chiba 277-8581, Japan
Email: fukuyama@issp.u-tokyo.ac.jp
http://www.issp.u-tokyo.ac.jp/index_e.html

Andrei Ruckenstein
Editor for The Americas

Department of Physics and Astronomy
Rutgers, The State University of New Jersey
136 Frelinghuysen Road
Piscataway, NJ 08854-8019, USA
Phone: +1 (732) 445 43 29
Fax: +1 (732) 445-43 43
Email: andreir@physics.rutgers.edu
http://www.physics.rutgers.edu/people/pips/Ruckenstein.html

Peter Wölfle

Institut für Theorie der Kondensierten Materie
Universität Karlsruhe
Postfach 69 80
76128 Karlsruhe, Germany
Phone: +49 (7 21) 6 08 35 90
Fax: +49 (7 21) 69 81 50
Email: woelfle@tkm.physik.uni-karlsruhe.de
http://www-tkm.physik.uni-karlsruhe.de

Complex Systems, Editor

Frank Steiner

Abteilung Theoretische Physik
Universität Ulm
Albert-Einstein-Allee 11
89069 Ulm, Germany
Phone: +49 (7 31) 5 02 29 10
Fax: +49 (7 31) 5 02 29 24
Email: steiner@physik.uni-ulm.de
http://www.physik.uni-ulm.de/theo/theophys.html

Maurizio Dapor

Electron-Beam Interactions with Solids

Application of the Monte Carlo Method to Electron Scattering Problems

With 27 Figures

 Springer

Dr. Maurizio Dapor
Centro per la ricerca scientifica e tecnologica
Via Sommarive 18
38050 Povo (Trento), Italy
E-mail: dapor@itc.it

Cataloging-in-Publication Data applied for

A catalog record for this book is available from the Library of Congress.

Bibliographic information published by Die Deutsche Bibliothek

Die Deutsche Bibliothek lists this publication in the Deutsche Nationalbibliografie; detailed bibliographic data is available in the Internet at http://dnb.ddb.de.

Physics and Astronomy Classification Scheme (PACS):
05.10.Ln, 34.80-i, 34.80.Bm, 34.80.Dp, 68.49.Jk

ISSN print edition: 0081-3869
ISSN electronic edition: 1615-0430
ISBN 3-540-00652-4 Springer-Verlag Berlin Heidelberg New York

This work is subject to copyright. All rights are reserved, whether the whole or part of the material is concerned, specifically the rights of translation, reprinting, reuse of illustrations, recitation, broadcasting, reproduction on microfilm or in any other way, and storage in data banks. Duplication of this publication or parts thereof is permitted only under the provisions of the German Copyright Law of September 9, 1965, in its current version, and permission for use must always be obtained from Springer-Verlag. Violations are liable for prosecution under the German Copyright Law.

Springer-Verlag Berlin Heidelberg New York
a member of BertelsmannSpringer Science+Business Media GmbH

http://www.springer.de

© Springer-Verlag Berlin Heidelberg 2003
Printed in Germany

The use of general descriptive names, registered names, trademarks, etc. in this publication does not imply, even in the absence of a specific statement, that such names are exempt from the relevant protective laws and regulations and therefore free for general use.

Typesetting: Author and LE-TEX GbR, Leipzig using a Springer LATEX macro package
Cover concept: eStudio Calamar Steinen
Cover production: *design & production* GmbH, Heidelberg

Printed on acid-free paper   57/3141/YL   5 4 3 2 1 0

To my mother and father

Preface

The interaction of an electron beam with a solid target has been studied since the early part of the past century. Since 1960, the electron–solid interaction has become the subject of a number of investigators' work owing to its fundamental role in scanning electron microscopy, in electron-probe microanalysis, in Auger electron spectroscopy, in electron-beam lithography and in radiation damage. The interaction of an electron beam with a solid target has often been investigated theoretically by using the Monte Carlo method, a numerical procedure involving random numbers that is able to solve mathematical problems. This method is very useful for the study of electron penetration in matter. The probabilistic laws of the interaction of an individual electron with the atoms constituting the target are well known. Consequently, it is possible to compute the macroscopic characteristics of interaction processes by simulating a large number of real trajectories, and then averaging them. The aim of this book is to study the probabilistic laws of the interaction of individual electrons with atoms (elastic and inelastic cross-sections); to investigate selected aspects of electron interaction with matter (backscattering coefficients for bulk targets, absorption, backscattering and transmission for both supported and unsupported thin films, implantation profiles, secondary-electron emission, and so on); and to introduce the Monte Carlo method and its applications to compute the macroscopic characteristics of the interaction processes mentioned above. The book compares theory, computational simulations and experimental data in order to offer a more global vision.

I wish to thank Antonio Miotello (Trento University) and Francesc Salvat (Barcelona University) for their invaluable and stimulating comments. I am also very pleased to acknowledge Angela Lahee (Springer-Verlag) for her constructive cooperation. I appreciate the conscientiousness of Douglas Meekison (Springer-Verlag) who read the entire manuscript and bettered it considerably. I am indebted to Hayley Stead for her skilful technical assistance in checking and improving the quality of the English of the book. Warm thanks are due to my wife and children for their immense patience and understanding during the period when this monograph was written.

Povo, January 2003 *Maurizio Dapor*

Contents

1. **Introduction** 1
   1.1 Motivation 1
       1.1.1 Backscattered Particles 2
       1.1.2 Secondary Electrons 2
       1.1.3 Depth Distribution of Trapped Particles 2
   1.2 From the Dirac Equation to the Monte Carlo Simulation 3
   References 4

2. **The Spin of the Electron** 5
   2.1 The Spectrum of Angular Momentum 5
   2.2 The Spin of the Electron 7
   2.3 The Dirac Equation 9
   2.4 The Solution for Free Particles 11
   2.5 The Dirac Equation in a Central Potential 13
   References 16

3. **Elastic Scattering** 17
   3.1 The First Born Approximation 17
   3.2 The Density Matrix and Spin Polarisation 22
   3.3 Relativistic Partial-Wave Expansion 25
   3.4 Calculation of the Phase Shifts 34
   3.5 Exchange and Solid State Effects 37
   3.6 Comparing Theory and Experimental Data 37
   References 41

4. **Inelastic Scattering** 43
   4.1 The Classical Theory 43
   4.2 Dielectric Function and Stopping Power 45
   4.3 Inelastic Mean Free Path 46
   4.4 Positrons 47
   4.5 Plasma Oscillations 47
   4.6 Comparing Theory and Experimental Data 48
   References 52

X   Contents

5. **Electrons Impinging on Solid Targets** 53
   5.1 Backscattered Electrons 53
   5.2 Electrons in thin films 56
       5.2.1 Definitions, Symbols, Properties 56
       5.2.2 Unsupported thin films 58
       5.2.3 Supported thin films 59
   5.3 Secondary Electrons 61
   5.4 Comparing Theory and Experimental Data 65
   References 68

6. **Monte Carlo Simulations** 69
   6.1 The Monte Carlo Method 69
   6.2 Random Variables 70
       6.2.1 Random Variable Uniformly Distributed
             in the Interval (0, 1) 71
       6.2.2 Random Variable Distributed
             in a Given Interval with a Given Probability 71
       6.2.3 Random Variable Uniformly Distributed
             in the Interval (a, b) 72
       6.2.4 Random Variable with Poisson Distribution 72
       6.2.5 Pseudo-Random-Number Generators 73
   6.3 A Simple Monte Carlo Scheme 73
   6.4 A More Sophisticated Simulation 76
       6.4.1 Surface Films 77
   6.5 Another Monte Carlo Scheme 78
       6.5.1 Angular Deflection in Electron–Electron Collisions. 79
       6.5.2 Secondary Electrons 80
   6.6 Comparing Theory and Experimental Data 81
   References 90

A. **Matrices and Operators** 91
   A.1 Representation of Linear Operators 91
   A.2 Matrix Transformations 91
   A.3 Commuting Operators 93

B. **The Dirac Notation** 95
   B.1 Ket and Bra Vectors 95
   B.2 Continuous Spectrum 96
   B.3 The Schrödinger Equation in the Dirac Notation 98

C. **Special Functions** 99
   C.1 Legendre Polynomials
       and Associated Legendre Functions 99
   C.2 Bessel Functions 100
   C.3 The Spherical Harmonics 102

Index 105

1 Introduction

1.1 Motivation

Studies of low-energy backscattered electrons, secondary electrons and absorbed electron depth distribution, have all been of great interest for many analytical techniques. Indeed, a better understanding of the collision events occurring in the surface layers before the emission of backscattered and secondary electrons, should allow a more general comprehension of surface physics. The problem of backscattered and secondary electron emission from solids irradiated by a particle beam is of crucial importance, especially in connection with the analytical techniques that utilise secondary electrons to investigate chemical and compositional properties of solids in the near surface layers, i.e. Auger electron spectroscopy and X-photoelectron spectroscopy. The electron backscattering coefficient is a quantity which, for slow primary electrons and for supported and unsupported thin films, requires theoretical, numerical and experimental investigation. The energy spectra of secondary electrons emitted by solid targets and stimulated by particle beams are very complicated because many features appear in the spectra due to the different collision processes involved prior to low-energy secondary electron emission. Accurate knowledge of the depth distribution of trapped electrons in dielectric materials (biological materials, ceramics, glass) is necessary to solve the diffusion equation in order to investigate the charging phenomena, while the study of positron depth distribution in solids is required for positron annihilation spectroscopy, a technique that allows non-destructive investigations of structural defects at surfaces and interfaces.

As a consequence, the backscattered electrons and positrons, the secondary electrons, and the depth distribution of the electrons and positrons absorbed by solid targets, all need to be theoretically and numerically investigated.

These, and other related quantities, can be accurately calculated using Monte Carlo simulations.

This book aims to provide the reader with the physical and mathematical instruments necessary to realise a Monte Carlo code in order to simulate the quantities quoted. But before proceeding, we would like to briefly add some words about the physical quantities just introduced.

1.1.1 Backscattered Particles

When an electron or a positron beam impinges on a solid, some particles of the beam scatter back and emerge from the target surface. These particles are known as backscattered electrons. Because of their penetration below the surface and the resulting loss of small amounts of energy through ionizations, electron excitations and plasmon emissions, the backscattered electrons are not reflected without dissipation of energy. A fraction of their energy is indeed lost into the solid before emerging.

When the target is a surface film (i.e. a thin film deposited on a substrate of a different material), the number of backscattered electrons depends on the film thickness, on the electron primary energy, and on the materials constituting the film and the substrate.

In simulating the processes of penetration and backscattering of electrons impinging on supported thin films with the Monte Carlo method, one has to take into account not only the interface between the film and the substrate but also the change in the scattering probabilities per unit length in passing from the film to the substrate and vice versa.

The fractions of electrons backscattered from unsupported thin films, solid targets, and surface films of various thickness and deposited on different substrates, can be accurately calculated using Monte Carlo simulations.

1.1.2 Secondary Electrons

The concepts which form the basis for a theoretical understanding of the problem of secondary electron emission have been analysed in Wolff's theory [1]. However, such a theory must include some simplification which may be valid in only a limited number of situations in order to attain analytical results.

To overcome these limitations, we have therefore set out, in this book, to describe a Monte Carlo procedure to accurately calculate the number of secondary electrons emitted from a solid irradiated with positron or electron beams.

1.1.3 Depth Distribution of Trapped Particles

When an insulator is subject to electron irradiation, the injected electrons cannot be definitely trapped but must instead recombine with positive charges left near the irradiated surface when secondary electrons are emitted: this is justified on the basis that dielectric breakdown is not observed during specific experiments of electron-irradiation of insulators. The dynamics of the absorbed electrons depend on a number of parameters: the fraction of trapped electrons, the space-charge distribution, the mobility, and the number of secondary electrons emitted from the region near the surface of the dielectric. The time evolution of the surface electric field can be studied by integrating the continuity equation for the relevant transport processes of the injected charge by adopting, as the charge source term, the distribution of the absorbed electrons as obtained by a Monte Carlo simulation.

The study of the interaction of positron beams with solid targets has been approached by various researchers, due to its importance for positron annihilation spectroscopy. This technique allows non-destructive investigations of the structural defects of surfaces and interfaces: in particular, information is provided about the nature and distribution of point defects in solid materials. The solution of the diffusion equation (necessary to obtain the fractions of incident positrons annihilated at different depths within the target) requires knowledge of the positron stopping profile, i.e., the initial depth distribution of the thermalized positrons that can be calculated through Monte Carlo simulation. We also note that the transmission of positrons is of great interest because, once backscattering is known, it allows calculation of the total fraction of particles absorbed by the target as a function of depth and primary energy. The calculation of the transmission of particles through thin films is also a typical application of Monte Carlo simulations and will be described in the last chapter.

1.2 From the Dirac Equation to the Monte Carlo Simulation

In the previous section, several examples of physical problems concerning the interaction of electrons and/or positrons with solids have been given. We have also noted that these problems are often solved by using Monte Carlo simulation, a computational tool based on the use of random variables.

An accurate knowledge of the elastic and inelastic cross-sections is needed to study the electron and positron penetration in solids.

The calculation of the scattering cross-sections, necessary for the Monte Carlo simulations, requires the study of the electron spin and the Dirac equation. This important equation will be described in Chap. 2. The following chapters will be dedicated to the problem of calculating cross-sections and to the description of the physical quantities involved in the electron–solid interaction that we are interested in investigating. The last chapter describes the Monte Carlo codes.

Numerical procedures to calculate the differential elastic scattering cross-sections necessary in the Monte Carlo simulations are described in Chap. 3.

The codes for the calculation of the elastic scattering cross-section are based on the quantum-relativistic partial-wave expansion method. The phase shifts (necessary to compute the differential elastic scattering cross-section) can be calculated by numerically solving the Dirac equation for a central electrostatic field up to a large radius in which the atomic potential is negligible. The atomic potential used here is that of Hartree–Fock for atomic numbers smaller than or equal to 18, and that of Dirac–Hartree–Fock–Slater for atomic numbers greater than 18.

The inelastic scattering can be quantified by using the inelastic mean free path and the stopping power, a quantity related to the probability of energy loss per unit distance travelled by the electron within the solid. An electron can lose a large fraction of its energy in a single collision: nevertheless the continuous slowing down approximation is generally accepted. In this approximation, the electron is assumed to continuously dissipate its energy whilst travelling within the solid. In this book, the inelastic mean free path and the stopping power are described following the method proposed by Ashley [2, 3]. Numerical procedures to calculate the inelastic mean free path and the stopping power, necessary in the Monte Carlo simulations, are described in Chap. 4.

Selected aspects of electron interaction with matter (backscattering coefficients for bulk targets, absorption, backscattering and transmission for both supported and unsupported thin films, implantation profiles, secondary-electron emission, and so on) are described in Chap. 5.

The Monte Carlo codes described in Chap. 6 give results which agree well with the experimental data in the electron and positron energy range from 1 000 to 30 000 eV. The numerical and theoretical results presented in this book have been utilised by several authors in their simulations (see, for example, [4, 5, 6]).

The mathematical and physical requirement are kept within the topics covered by a calculus course and by an introductory quantum mechanics course, respectively. The appendix provides some useful tools, i.e. a summary of the theory of matrices and operators, the Dirac notation (ket and bra vectors) and a description of the special functions of the mathematical physics (Legendre polynomials, associated Legendre functions, Bessel functions and spherical harmonics).

References

1. P.A. Wolff, Phys. Rev. **95**, 56 (1954)
2. J.C. Ashley, J. Electron Spectrosc. Relat. Phenom. **46**, 199 (1988)
3. J.C. Ashley, J. Electron Spectrosc. Relat. Phenom. **50**, 323 (1990)
4. C.L. Lee, K.Y. Kong, H. Gong, C.K. Ong, Surf. Interface Anal. **24**, 15 (1996)
5. Z. Chaoui, N. Bouarissa, Phys. Lett. A **297**, 432 (2002)
6. N. Bouarissa, B. Deghfel, A. Bentabet, Eur. Phys. J.: Appl. Phys. **19**, 89 (2002)

2 The Spin of the Electron

2.1 The Spectrum of Angular Momentum

Let us introduce the angular momentum operator in the treatment of single particle quantum systems [1, 2, 3, 4]. Let us indicate the electron mass by m, the electron position by $\boldsymbol{r}$, the electron energy by E and the electron momentum by $\boldsymbol{p}$. In quantum mechanics we assume the following correspondence rules relating the differential operators (Appendix A) and the physical quantities:

$$E \to i\hbar \frac{\partial}{\partial t}, \tag{2.1}$$

$$\boldsymbol{p} \to -i\hbar \boldsymbol{\nabla}. \tag{2.2}$$

Here $\hbar = h/2\pi$ and $h = 4.136 \times 10^{-15}$ eV sec is the Planck constant. The differential operators act on wave functions that are square-integrable complex functions in a Hilbert space. Now, if we consider the components of the electron orbital angular momentum $\boldsymbol{L} = \boldsymbol{r} \times \boldsymbol{p}$, using the definition of $\boldsymbol{L}$ it is possible to see that

$$[L_x, L_y] = i\hbar L_z, \tag{2.3}$$

$$[L_y, L_z] = i\hbar L_x, \tag{2.4}$$

$$[L_z, L_x] = i\hbar L_y. \tag{2.5}$$

The commutator is defined for any pair of linear operators A and B as

$$[A, B] = AB - BA. \tag{2.6}$$

In order to introduce the intrinsic angular momentum, we can begin by generalising the properties of $\boldsymbol{L}$, saying that a linear operator $\boldsymbol{J}$ is an angular momentum (orbital or intrinsic) if

$$[J_x, J_y] = i\hbar J_z, \tag{2.7}$$

$$[J_y, J_z] = i\hbar J_x, \tag{2.8}$$

$$[J_z, J_x] = i\hbar J_y. \tag{2.9}$$

As the reader may easily verify, the two operators $\boldsymbol{J}^2$ and J_z commute, i.e.

$$[J_z, \boldsymbol{J}^2] = 0 , \tag{2.10}$$

so they possess at least one basis of eigenvectors in common. As a consequence, the corresponding physical quantities may be measured together and simultaneously with an arbitrary precision. Using the Dirac notation (Appendix B) and using $|jm\rangle$ to indicate a basis of orthonormal eigenvectors that $\boldsymbol{J}^2$ and J_z have in common, the eigenvalues $\hbar^2 j(j+1)$ and $\hbar m$ correspond respectively to $\boldsymbol{J}^2$ and J_z, whereby

$$\boldsymbol{J}^2 |jm\rangle = \hbar^2 j(j+1)|jm\rangle , \tag{2.11}$$

$$J_z |jm\rangle = \hbar m |jm\rangle . \tag{2.12}$$

In order to proceed, let us now define the two operators J_- and J_+ by

$$J_\pm = J_x \pm i J_y , \tag{2.13}$$

which have the following properties:

$$J_\pm^\dagger = J_\mp , \tag{2.14}$$

$$[J_z, J_\pm] = \pm \hbar J_\pm , \tag{2.15}$$

$$J_+ J_- = J_x^2 + J_y^2 + \hbar J_z , \tag{2.16}$$

$$J_- J_+ = J_x^2 + J_y^2 - \hbar J_z , \tag{2.17}$$

$$[J_+, J_-] = 2\hbar J_z , \tag{2.18}$$

$$\{J_+, J_-\} = 2(J_x^2 + J_y^2) . \tag{2.19}$$

Here we have introduced the anticommutator, defined for any pair of linear operators A and B by

$$\{A, B\} = AB + BA . \tag{2.20}$$

As a consequence,

$$\boldsymbol{J}^2 = J_z^2 + \frac{1}{2}(J_+ J_- + J_- J_+) . \tag{2.21}$$

Let us now calculate the norms of the vectors $J_- |jm\rangle$ and $J_+ |jm\rangle$, i.e.

$$\langle jm| J_-^\dagger J_- |jm\rangle = \langle jm| J_+ J_- |jm\rangle = \hbar^2 [j(j+1) - m(m-1)] \geq 0 , \tag{2.22}$$

$$\langle jm| J_+^\dagger J_+ |jm\rangle = \langle jm| J_- J_+ |jm\rangle = \hbar^2 [j(j+1) - m(m+1)] \geq 0 . \tag{2.23}$$

As the norms are positive,

$$-j \leq m \leq j . \tag{2.24}$$

Note that $m = -j$ if and only if $J_-|jm\rangle = 0$, while $m = j$ if and only if $J_+|jm\rangle = 0$. As $\boldsymbol{J}^2$ commutes with $J_\pm$, then

$$\boldsymbol{J}^2 J_\pm|jm\rangle = J_\pm \boldsymbol{J}^2|jm\rangle = \hbar^2 j(j+1) J_\pm|jm\rangle . \tag{2.25}$$

From (2.15), we obtain

$$J_z J_\pm|jm\rangle = J_\pm J_z|jm\rangle \pm \hbar J_\pm|jm\rangle = \hbar(m \pm 1) J_\pm|jm\rangle . \tag{2.26}$$

The last two equations tell us that the vectors $J_\pm|jm\rangle$ are common eigenvectors of $\boldsymbol{J}^2$ and J_z. On the other hand, comparing the last equation with

$$J_z|jm \pm 1\rangle = \hbar(m \pm 1)|jm \pm 1\rangle , \tag{2.27}$$

we can see that the eigenvectors $J_\pm|jm\rangle$ must be proportional to $|jm \pm 1\rangle$. Therefore, taking into account the values of their norms that we have just calculated, we can write that

$$J_\pm|jm\rangle = \hbar\sqrt{j(j+1) - m(m \pm 1)}|jm \pm 1\rangle . \tag{2.28}$$

Following the last equations, simple considerations about the integer number of steps p necessary to go from $m = -j$ to $m = j$ using the operator J_+ ($-j + p = j$) allow us to conclude that j must be an integral or half-integral non-negative number,

$$j = 0, \frac{1}{2}, 1, \frac{3}{2}, 2, \frac{5}{2}, \ldots, \infty , \tag{2.29}$$

and that the only possibles values of m are integral and half-integral numbers

$$m = 0, \pm\frac{1}{2}, \pm 1, \pm\frac{3}{2}, \pm 2, \pm\frac{5}{2}, \ldots, \pm\infty . \tag{2.30}$$

In conclusion, if $\hbar j(j+1)$ and $\hbar m$ are, respectively, the eigenvalues of $\boldsymbol{J}^2$ and J_z then j must be an integral or half-integral non-negative quantity, and the only possible values of m for any given j are the $2j + 1$ numbers

$$-j, -j+1, \ldots, j-1, j . \tag{2.31}$$

2.2 The Spin of the Electron

As we are interested in electrons and in their intrinsic angular momentum, or spin, we shall now focus our attention on the case $j = 1/2$. Let us simplify the equations by introducing the following notation:

$$|-\rangle \equiv |1/2 \ -1/2\rangle , \tag{2.32}$$

$$|+\rangle \equiv |1/2 \ +1/2\rangle . \tag{2.33}$$

Let us indicate the spin operator by $\boldsymbol{S}$ and introduce the two-dimensional spin eigenspace defined by the two eigenvectors $|-\rangle$ and $|+\rangle$, possessing the following properties:

2 The Spin of the Electron

$$\langle +|-\rangle = \langle -|+\rangle = 0 , \tag{2.34}$$

$$\langle -|-\rangle = \langle +|+\rangle = 1 , \tag{2.35}$$

$$\boldsymbol{S}^2|-\rangle = \frac{3\hbar^2}{4}|-\rangle , \tag{2.36}$$

$$\boldsymbol{S}^2|+\rangle = \frac{3\hbar^2}{4}|+\rangle , \tag{2.37}$$

$$S_z|-\rangle = -\frac{\hbar}{2}|-\rangle , \tag{2.38}$$

$$S_z|+\rangle = \frac{\hbar}{2}|+\rangle . \tag{2.39}$$

The general case of a spin-1/2 state ξ is a linear superimposition of the basic eigenvectors $|-\rangle$ and $|+\rangle$, whereby

$$\xi = A|+\rangle + B|-\rangle , \tag{2.40}$$

where the coefficients A and B are complex numbers. $|A|^2$ is the probability of finding the electron in the state of "spin up" along the z axis; $|B|^2$ is the probability of finding the electron in the "spin down" state. It follows that, from the condition of normalisation of ξ,

$$|A|^2 + |B|^2 = 1 . \tag{2.41}$$

Using the equations

$$S_-|-\rangle = 0 , \tag{2.42}$$

$$S_-|+\rangle = \hbar|-\rangle , \tag{2.43}$$

$$S_+|-\rangle = \hbar|+\rangle , \tag{2.44}$$

$$S_+|+\rangle = 0 , \tag{2.45}$$

we can calculate the matrix elements of S_+ and S_-, as follows:

$$S_+ = \hbar \begin{pmatrix} 0 & 1 \\ 0 & 0 \end{pmatrix} , \tag{2.46}$$

$$S_- = \hbar \begin{pmatrix} 0 & 0 \\ 1 & 0 \end{pmatrix} . \tag{2.47}$$

Let us now define the Pauli matrices σ_z, σ_y and σ_z by

$$\boldsymbol{S} = \frac{\hbar}{2}\boldsymbol{\sigma}. \tag{2.48}$$

The representation of the Pauli matrices in the basis $\{|-\rangle, |+\rangle\}$ can be easily obtained, observing that

$$S_x = \frac{1}{2}(S_- + S_+),\qquad(2.49)$$

$$S_y = \frac{i}{2}(S_- - S_+).\qquad(2.50)$$

The Pauli matrices in the basis $\{|-\rangle,|+\rangle\}$ are

$$\sigma_x = \begin{pmatrix} 0 & 1 \\ 1 & 0 \end{pmatrix},\qquad(2.51)$$

$$\sigma_y = \begin{pmatrix} 0 & -i \\ i & 0 \end{pmatrix},\qquad(2.52)$$

$$\sigma_z = \begin{pmatrix} 1 & 0 \\ 0 & -1 \end{pmatrix}.\qquad(2.53)$$

Using this particular representation or their definition, we can see that the Pauli matrices are Hermitian and have the following properties:

$$\sigma_x^2 = \sigma_y^2 = \sigma_z^2 = 1,\qquad(2.54)$$
$$\sigma_x \sigma_y \sigma_z = i,\qquad(2.55)$$
$$\sigma_x \sigma_y - \sigma_y \sigma_x = 2i\sigma_z,\qquad(2.56)$$
$$\sigma_z \sigma_x - \sigma_x \sigma_z = 2i\sigma_y,\qquad(2.57)$$
$$\sigma_y \sigma_z - \sigma_z \sigma_y = 2i\sigma_x,\qquad(2.58)$$
$$\sigma_x \sigma_y + \sigma_y \sigma_x = 0,\qquad(2.59)$$
$$\sigma_z \sigma_x + \sigma_x \sigma_z = 0,\qquad(2.60)$$
$$\sigma_y \sigma_z + \sigma_z \sigma_y = 0,\qquad(2.61)$$
$$\operatorname{tr} \sigma_x = \operatorname{tr} \sigma_y = \operatorname{tr} \sigma_z = 0,\qquad(2.62)$$
$$\det \sigma_x = \det \sigma_y = \det \sigma_z = -1.\qquad(2.63)$$

2.3 The Dirac Equation

The energy of a non-relativistic particle of momentum $\boldsymbol{p}$ and mass m in a potential $V(\boldsymbol{r})$ is given by

$$E = \frac{\boldsymbol{p}^2}{2m} + V(\boldsymbol{r}).\qquad(2.64)$$

The correspondence rules relating the differential operators and the physical quantities are

$$E \to i\hbar \frac{\partial}{\partial t},\qquad(2.65)$$

$$\boldsymbol{p} \to -i\hbar \boldsymbol{\nabla}.\qquad(2.66)$$

When applied to the energy of the particle they give the non-relativistic Hamiltonian operator of a particle in a field,

$$H = -\frac{\hbar^2}{2m}\nabla^2 + V(\boldsymbol{r}) \,, \tag{2.67}$$

and the Schrödinger equation describing the time evolution of the wave function,

$$i\hbar\frac{\partial}{\partial t}\psi = \left[-\frac{\hbar^2}{2m}\nabla^2 + V(\boldsymbol{r})\right]\psi \,, \tag{2.68}$$

or

$$i\hbar\frac{\partial}{\partial t}\psi = H\psi \,. \tag{2.69}$$

The Dirac equation is the appropriate equation to treat the electron spin and relativistic effects. The square of the energy E of a relativistic free electron is given by

$$E^2 = \boldsymbol{p}^2 c^2 + m^2 c^4 \,, \tag{2.70}$$

where c is the speed of light, m is the rest mass of the electron and $\boldsymbol{p}$ is the electron momentum. In the following, all equations will be written with a choice of units such that

$$\hbar = c = 1 \,, \tag{2.71}$$

in which time has the dimensions of length, the electron charge e is a dimensionless quantity such that

$$e^2 \approx \frac{1}{137} \,, \tag{2.72}$$

and the square of the energy of a relativistic free electron becomes

$$E^2 = \boldsymbol{p}^2 + m^2 \,. \tag{2.73}$$

Note that through homogeneity considerations, we can easily re-establish the general equations.

For the free electron, the Dirac Hamiltonian is given by

$$H = \boldsymbol{\alpha} \cdot \boldsymbol{p} + \beta m = -i\boldsymbol{\alpha} \cdot \nabla + \beta m \,, \tag{2.74}$$

in which $\boldsymbol{\alpha} = (\alpha_x, \alpha_y, \alpha_z) = (\alpha^1, \alpha^2, \alpha^3)$ and β are four Hermitian operators. Therefore,

$$i\frac{\partial}{\partial t}\phi = [-i\boldsymbol{\alpha} \cdot \nabla + \beta m]\phi \,, \tag{2.75}$$

where ϕ is the wave function. The square of the Dirac Hamiltonian must obviously be equal to $\boldsymbol{p}^2 + m^2$:

$$(\alpha_x p_x + \alpha_y p_y + \alpha_z p_z + \beta m)^2 = \boldsymbol{p}^2 + m^2, \tag{2.76}$$

and therefore

$$\sum_j \alpha^{j2} p^{j2} + \beta^2 m^2 + \sum_{j<k}(\alpha^j \alpha^k + \alpha^k \alpha^j) p^j p^k + \sum_j (\alpha^j \beta + \beta \alpha^j) p^j$$

$$= m^2 + \sum_j p^{j2} . \quad (2.77)$$

As a consequence, the four operators α^j and β must satisfy the following relations:

$$\alpha^{j2} = \beta^2 = 1 ,$$
$$\alpha^j \alpha^k + \alpha^k \alpha^j = 0 \text{ for } j \neq k ,$$
$$\alpha^j \beta + \beta \alpha^j = 0 . \quad (2.78)$$

As they must obey these relations, the operators α_i and β obviously cannot be numbers; instead they must be Hermitian matrices operating on a multi-component column vector ϕ (which we call a spinor). The lowest-dimensionality matrices that obey the relations (2.78) are 4×4. If $(\sigma_x, \sigma_y, \sigma_z) = (\sigma^1, \sigma^2, \sigma^3)$ are the 2×2 Pauli matrices, and I the 2×2 identity matrix, then the following 4×4 matrices satisfy the conditions expressed by (2.78):

$$\alpha^j = \begin{pmatrix} 0 & \sigma^j \\ \sigma^j & 0 \end{pmatrix} , \quad (2.79)$$

$$\beta = \begin{pmatrix} I & 0 \\ 0 & -I \end{pmatrix} . \quad (2.80)$$

In the presence of an electromagnetic field described by the four-potential $(\varphi, \boldsymbol{A})$, the Dirac Hamiltonian becomes ($E \to E - e\varphi$, $\boldsymbol{p} \to \boldsymbol{p} - e\boldsymbol{A}$)

$$H = e\varphi + \boldsymbol{\alpha} \cdot (\boldsymbol{p} - e\boldsymbol{A}) + \beta m , \quad (2.81)$$

so that the Dirac equation describing a spin-1/2 particle in the presence of an electromagnetic field is

$$\left[\left(i\frac{\partial}{\partial t} - e\varphi\right) - \boldsymbol{\alpha} \cdot (-i\boldsymbol{\nabla} - e\boldsymbol{A}) - \beta m\right]\phi = 0 , \quad (2.82)$$

where $\phi(\boldsymbol{r}, t)$ is a four-component spinor depending on $\boldsymbol{r}$ and on t.

2.4 The Solution for Free Particles

In the case of free electrons, the Hamiltonian does not depend on $\boldsymbol{r}$ and t, so that we can express ϕ as a plane wave:

$$\phi = u(\boldsymbol{p}) \exp[i(\boldsymbol{p} \cdot \boldsymbol{r} - Et)] . \quad (2.83)$$

2 The Spin of the Electron

As ϕ must obey the free Dirac equation, then the function $u(\boldsymbol{p})$ must be a four-component spinor independent of $\boldsymbol{r}$ that satisfies the equation

$$(\boldsymbol{\alpha} \cdot \boldsymbol{p} + \beta m)u(\boldsymbol{p}) = Eu(\boldsymbol{p}) . \tag{2.84}$$

This is equivalent to

$$\begin{cases} (E-m)u_1 & -p_z u_3 -(p_x - ip_y)u_4 = 0 \\ (E-m)u_2 -(p_x + ip_y)u_3 & +p_z u_4 = 0 \\ -p_z u_1 -(p_x - ip_y)u_2 +(E+m)u_3 & = 0 \\ -(p_x + ip_y)u_1 +p_z u_2 & +(E+m)u_4 = 0 \end{cases}$$

or, by choosing the z axis in the direction of $\boldsymbol{p}$ ($p_x = p_y = 0, p_z = p$),

$$\begin{cases} (E-m)u_1 & -pu_3 & = 0 \\ (E-m)u_2 & +pu_4 = 0 \\ -pu_1 & +(E+m)u_3 & = 0 \\ +pu_2 & +(E+m)u_4 = 0 \end{cases}$$

In order to find the eigenvalues of the energy, we have to solve the following equation:

$$\begin{vmatrix} E-m & 0 & -p & 0 \\ 0 & E-m & 0 & p \\ -p & 0 & E+m & 0 \\ 0 & p & 0 & E+m \end{vmatrix} = 0 .$$

Upon solving this last equation, we can deduce the following doubly degenerate eigenvalues for the energy of the free Dirac particle:

$$E_{\pm} = \pm\sqrt{p^2 + m^2} . \tag{2.85}$$

Note that the spin operator in the direction of $\boldsymbol{p}$ is the operator $\sigma_z/2 = -i\alpha_x\alpha_y/2$, which commutes with the Hamiltonian $H = \alpha_z p + \beta m$, so that the solutions we are seeking correspond to the eigenvectors which are common to H and $\sigma_z/2$. In order to find the first eigenvector, let us just impose $u_1 = 1$ and $u_2 = 0$ and normalise to unity ($u^\dagger u = 1$). In such a way, we find that

$$u_{\uparrow E_+}(p) = \sqrt{\frac{E_+ + m}{2E_+}} \begin{pmatrix} 1 \\ 0 \\ p/(E_+ + m) \\ 0 \end{pmatrix} , \tag{2.86}$$

where $u_{\uparrow E_+}(p)$ is the eigensolution corresponding to a particle with spin up and positive energy (i.e. a spin-up electron). The other eigenvectors can be obtained by proceeding in a similar way. The solution corresponding to a spin-down electron is

$$u_{\downarrow E_+}(p) = \sqrt{\frac{E_+ + m}{2E_+}} \begin{pmatrix} 0 \\ 1 \\ 0 \\ -p/(E_+ + m) \end{pmatrix} , \tag{2.87}$$

while the two spinors corresponding to the positron (negative energy) are

$$u_{\uparrow E_-}(p) = \sqrt{\frac{E_+ + m}{2E_+}} \begin{pmatrix} -p/(E_+ + m) \\ 0 \\ 1 \\ 0 \end{pmatrix}, \qquad (2.88)$$

$$u_{\downarrow E_-}(p) = \sqrt{\frac{E_+ + m}{2E_+}} \begin{pmatrix} 0 \\ p/(E_+ + m) \\ 0 \\ 1 \end{pmatrix}. \qquad (2.89)$$

Note that in the non-relativistic limit the solutions corresponding to the electron (positive energy) have the two components u_1 and u_2 much greater than u_3 and u_4 so that, for $v \ll c$, the four-component Dirac spinors reduce to the two-component spinors of the Pauli theory.

2.5 The Dirac Equation in a Central Potential

In order to appropriately treat the quantum-relativistic scattering theory, we need to know the form assumed by the Dirac equation for an electron in the presence of a central electrostatic field described by a central potential $e\varphi(r) = V(r)$.

Let us first introduce the operator $\mathcal{K}$ defined by

$$\mathcal{K} \equiv \beta(1 + \boldsymbol{\sigma} \cdot \boldsymbol{L}), \qquad (2.90)$$

where $\boldsymbol{L}$ is the electron orbital angular momentum. For an electron in a central electrostatic field, it is possible to show that

$$2\frac{d\boldsymbol{L}}{dt} = -\frac{d\boldsymbol{\sigma}}{dt}. \qquad (2.91)$$

As a consequence, the total angular momentum, defined as $\boldsymbol{J} = \boldsymbol{L} + (1/2)\boldsymbol{\sigma}$, is a constant of the motion. On the other hand,

$$\boldsymbol{J}^2 - \boldsymbol{L}^2 = \boldsymbol{\sigma} \cdot \boldsymbol{L} + \frac{3}{4}, \qquad (2.92)$$

so we can conclude that

$$\mathcal{K} = \beta(\boldsymbol{J}^2 - \boldsymbol{L}^2 + 1/4). \qquad (2.93)$$

$\mathcal{K}$ commutes with H and is, as a consequence, a constant of the motion.

Let us now define the radial-momentum operator p_r, where

$$p_r \equiv -i\frac{1}{r}\frac{\partial}{\partial r}r = \frac{\boldsymbol{r} \cdot \boldsymbol{p} - i}{r}, \qquad (2.94)$$

and introduce the radial component α_r of the α operator, where

$$\alpha_r = \frac{\boldsymbol{\alpha} \cdot \boldsymbol{r}}{r} ; \tag{2.95}$$

this obeys the relation

$$\alpha_r^2 = 1 . \tag{2.96}$$

For any pair of vectors $\boldsymbol{a}$ and $\boldsymbol{b}$, the following equations hold:

$$(\boldsymbol{\sigma} \cdot \boldsymbol{a})(\boldsymbol{\sigma} \cdot \boldsymbol{b}) = \boldsymbol{a} \cdot \boldsymbol{b} + i\boldsymbol{\sigma} \cdot \boldsymbol{a} \times \boldsymbol{b} \tag{2.97}$$

and

$$(\boldsymbol{\alpha} \cdot \boldsymbol{a})(\boldsymbol{\alpha} \cdot \boldsymbol{b}) = (\boldsymbol{\sigma} \cdot \boldsymbol{a})(\boldsymbol{\sigma} \cdot \boldsymbol{b}) . \tag{2.98}$$

As a consequence,

$$(\boldsymbol{\alpha} \cdot \boldsymbol{r})(\boldsymbol{\alpha} \cdot \boldsymbol{p}) = rp_r + i\beta\mathcal{K} . \tag{2.99}$$

This last equation is equivalent to

$$(\boldsymbol{\alpha} \cdot \boldsymbol{p}) = \alpha_r \left(p_r + \frac{i\beta\mathcal{K}}{r} \right) . \tag{2.100}$$

As a result, the Dirac equation with the Hamiltonian

$$H = \alpha_r \left(p_r + \frac{i\beta\mathcal{K}}{r} \right) + \beta m + V(r) \tag{2.101}$$

can be rewritten as the following:

$$\left[\alpha_r \left(p_r + \frac{i\beta\mathcal{K}}{r} \right) + \beta m + V(r) \right] \phi = E\phi . \tag{2.102}$$

The operators β, $\mathcal{K}$, L^2 and J_z are mutually commuting. In the following, $\mathcal{X}$ indicates an eigenvector common to those operators, so that

$$\beta\mathcal{X} = \mathcal{X} , \tag{2.103}$$

$$\mathcal{K}\mathcal{X} = -k\mathcal{X} , \tag{2.104}$$

$$L^2\mathcal{X} = l(l+1)\mathcal{X} , \tag{2.105}$$

$$J_z\mathcal{X} = m_j\mathcal{X} . \tag{2.106}$$

Introducing the function $\mathcal{Y}$, where

$$\mathcal{Y} = -\alpha_r\mathcal{X} , \tag{2.107}$$

we can observe that it has the following properties:

$$\mathcal{X} = -\alpha_r\mathcal{Y} , \tag{2.108}$$

$$\mathcal{K}\mathcal{Y} = -k\mathcal{Y} , \tag{2.109}$$

$$\beta\mathcal{Y} = \alpha_r\beta\mathcal{X} = -\mathcal{Y} . \tag{2.110}$$

2.5 The Dirac Equation in a Central Potential

Let us now consider the following linear combination of $\mathcal{X}$ and $\mathcal{Y}$:

$$\phi = F(r)\mathcal{Y} + iG(r)\mathcal{X} , \qquad (2.111)$$

which is an eigenvector common to H, $\mathcal{K}$ and J_z and thus the spinor we are looking for.

Our objective is to determine the functions $F(r)$ and $G(r)$. The eigenvalues of $\boldsymbol{L}^2$ are $(j\pm 1/2)(j\pm 1/2+1)$. As $\mathcal{K} = \beta(\boldsymbol{J}^2 - \boldsymbol{L}^2 + 1/4)$, the eigenvalues of $\mathcal{K}$ for the case $j = l + 1/2$ (spin up) are therefore given by

$$k = -\left(j + \frac{1}{2}\right) = -(l+1) . \qquad (2.112)$$

In the other case, where $j = l - 1/2$ (spin down), the eigenvalues of $\mathcal{K}$ are

$$k = \left(j + \frac{1}{2}\right) = l . \qquad (2.113)$$

We are now able to find the equations corresponding to the radial behaviour of the functions F and G. In order to do this, let us consider the Dirac equation (2.102) and observe the following:

$$\alpha_r p_r F(r)\mathcal{Y} = i\left[\frac{dF(r)}{dr} + \frac{F(r)}{r}\right]\mathcal{X} , \qquad (2.114)$$

$$i\alpha_r p_r G(r)\mathcal{X} = -\left[\frac{dG(r)}{dr} + \frac{G(r)}{r}\right]\mathcal{Y} , \qquad (2.115)$$

$$\frac{i\alpha_r \beta \mathcal{K}}{r} F(r)\mathcal{Y} = -\frac{i}{r}F(r)k\mathcal{X} , \qquad (2.116)$$

$$\frac{i\alpha_r \beta \mathcal{K}}{r} iG(r)\mathcal{X} = -\frac{1}{r}G(r)k\mathcal{Y} , \qquad (2.117)$$

$$\beta m F(r)\mathcal{Y} = -mF(r)\mathcal{Y} , \qquad (2.118)$$

$$\beta m iG(r)\mathcal{X} = imG(r)\mathcal{X} . \qquad (2.119)$$

Hence, since $\mathcal{X}$ and $\mathcal{Y}$ belong to different eigenvalues of β and are, consequently, linearly independent, the fundamental equations of the theory of the elastic scattering of electrons (and positrons) by atoms are

$$[E + m - V(r)]F(r) + \frac{dG(r)}{dr} + \frac{1+k}{r}G(r) = 0 , \qquad (2.120)$$

$$-[E - m - V(r)]G(r) + \frac{dF(r)}{dr} + \frac{1-k}{r}F(r) = 0 . \qquad (2.121)$$

References

1. A. Messiah, *Quantum Mechanics I and II* (North-Holland, Amsterdam, 1961)
2. H.A. Bethe, R. Jackiw, *Intermediate Quantum Mechanics* (Benjamin, New York, 1968)
3. F. Schwabl, *Quantum Mechanics* (Springer, Berlin, Heidelberg, 1992)
4. F. Schwabl, *Advanced Quantum Mechanics* (Springer, Berlin, Heidelberg, 1997)

3 Elastic Scattering

3.1 The First Born Approximation

Let us begin our investigation by considering the non-relativistic problem of electron–atom elastic scattering, i.e. the scattering of a beam of electrons by a central potential $V(r)$.

The solid angle $d\Omega$ depends on the scattering angles $[\theta, \theta + d\theta]$ and the azimuthal angles $[\phi, \phi + d\phi]$. The differential elastic scattering cross-section $d\sigma/d\Omega$ is defined as the ratio between the flux of particles per unit of time emerging in the solid angle $d\Omega$ (divided by $d\Omega$) and the incident flux.

The flux of particles per unit time emerging after the collision in the solid angle $d\Omega$ depends on the component j_r of the current density in the outgoing direction from the centre of the atomic nucleus. The number of electrons emerging in the solid angle $d\Omega$ per unit time is given by $j_r r^2 \, d\Omega$ (note that $r^2 \, d\Omega$ is the cross-sectional area normal to the radius).

Let us consider a beam of incident electrons in the direction z normalised to one particle per unit of volume. Let $K = mv/\hbar$ be the electron momentum in the z direction, where v is the electron velocity, m the electron mass and $\hbar$ the Planck constant divided by 2π. This beam can be represented by the plane wave $\exp(iKz)$.

Since the incident beam has been normalised to one particle per unit of volume, then the electron velocity v is the incident flux. As a consequence,

$$I(\theta,\phi) \equiv \frac{d\sigma}{d\Omega} = \frac{j_r r^2 \, d\Omega}{v \, d\Omega} = \frac{j_r r^2}{v} \ . \tag{3.1}$$

At a large distance from the atomic nucleus, the potential $V(r)$ is negligible and the scattered particles can be described by a spherical wave, i.e. a function which is proportional to $\exp(iKr)/r$. If $f(\theta, \phi)$ is the constant of proportionality (scattering amplitude), then the wave function $\Psi(\boldsymbol{r})$ of the whole scattering process (i.e. of the incident and the scattered electrons) satisfies the boundary conditions

$$\Psi(\boldsymbol{r}) \underset{r \to \infty}{\sim} \exp(iKz) + f(\theta,\phi)\frac{\exp(iKr)}{r} \ . \tag{3.2}$$

The electron position probability density P is given by $|\Psi|^2 = \Psi^*\Psi$, and the current density $\boldsymbol{j}(\boldsymbol{r},t)$ is

3 Elastic Scattering

$$j(r,t) = \frac{i\hbar}{2m}[(\boldsymbol{\nabla}\Psi^*)\Psi - \Psi^*(\boldsymbol{\nabla}\Psi)] \,. \tag{3.3}$$

Let us now calculate the radial component of the current density $\boldsymbol{j}$, j_r:

$$\begin{aligned} j_r &= \frac{i\hbar}{2m}\left\{f(\theta,\phi)\frac{\exp(iKr)}{r}\frac{\partial}{\partial r}\left[f^*(\theta,\phi)\frac{\exp(-iKr)}{r}\right]\right.\\ &\quad \left. - f^*(\theta,\phi)\frac{\exp(-iKr)}{r}\frac{\partial}{\partial r}\left[f(\theta,\phi)\frac{\exp(iKr)}{r}\right]\right\}\\ &= \frac{v|f(\theta,\phi)|^2}{r^2} \,. \end{aligned} \tag{3.4}$$

Comparing this equation with (3.1), we can see that the differential elastic scattering cross-section is the square of the modulus of $f(\theta,\phi)$:

$$I(\theta,\phi) = \frac{d\sigma}{d\Omega} = |f(\theta,\phi)|^2 \,. \tag{3.5}$$

In order to proceed, let us introduce the first Born approximation.

The first Born approximation is a high-energy approximation. If E is the incident electron energy, e the electron charge, a_0 the Bohr radius and Z the target atomic number, the first Born approximation is quite accurate if

$$E \gg \frac{e^2}{2a_0}Z^2 \,. \tag{3.6}$$

Let us now introduce the Green function and the integral-equation approach. Starting from the Schrödinger equation,

$$(\boldsymbol{\nabla}^2 + K^2)\Psi(\boldsymbol{r}) = \frac{2m}{\hbar^2}V(\boldsymbol{r})\Psi(\boldsymbol{r}) \,, \tag{3.7}$$

with the boundary condition expressed by (3.2), it is possible to show that this is a problem equivalent to the following integral equation:

$$\Psi(\boldsymbol{r}) = \exp(iKz) + \frac{2m}{\hbar^2}\int d^3r' g(\boldsymbol{r},\boldsymbol{r}')V(\boldsymbol{r}')\Psi(\boldsymbol{r}') \,, \tag{3.8}$$

in which

$$g(\boldsymbol{r},\boldsymbol{r}') = -\frac{\exp(iK|\boldsymbol{r}-\boldsymbol{r}'|)}{4\pi|\boldsymbol{r}-\boldsymbol{r}'|} \tag{3.9}$$

is the Green function of the operator $\boldsymbol{\nabla}^2 + K^2$. As is known, this operator satisfies the equation

$$(\boldsymbol{\nabla}^2 + K^2)g(\boldsymbol{r},\boldsymbol{r}') = \delta(\boldsymbol{r}-\boldsymbol{r}') \,, \tag{3.10}$$

where $\delta(\boldsymbol{r}-\boldsymbol{r}')$ is the Dirac delta function.

Let us apply the operator $\boldsymbol{\nabla}^2 + K^2$ to the function $\Psi(\boldsymbol{r})$ defined by the integral equation (3.8):

$$\begin{aligned}(\boldsymbol{\nabla}^2 + K^2)\Psi(\boldsymbol{r}) &= (\boldsymbol{\nabla}^2 + K^2)\exp(iKz)\\ &\quad + \frac{2m}{\hbar^2}\int d^3r'(\boldsymbol{\nabla}^2 + K^2)g(\boldsymbol{r},\boldsymbol{r}')V(\boldsymbol{r}')\Psi(\boldsymbol{r}') \,. \end{aligned}\tag{3.11}$$

3.1 The First Born Approximation

The application of the operator ∇^2 to the plane wave $\exp(iKz)$ gives

$$\nabla^2 \exp(iKz) = \frac{\partial^2}{\partial z^2}\exp(iKz) = -K^2 \exp(iKz) \tag{3.12}$$

and, as a consequence, we can write

$$(\nabla^2 + K^2)\exp(iKz) = 0 . \tag{3.13}$$

Therefore,

$$\begin{aligned}(\nabla^2 + K^2)\Psi(r) &= \frac{2m}{\hbar^2}\int d^3r' (\nabla^2 + K^2)g(r,r')V(r')\Psi(r')\\ &= \frac{2m}{\hbar^2}\int d^3r'\delta(r-r')V(r')\Psi(r')\\ &= \frac{2m}{\hbar^2}V(r)\Psi(r) .\end{aligned} \tag{3.14}$$

For the boundary conditions, we have

$$\begin{aligned}|r-r'| &= \sqrt{r^2 - 2r\cdot r' + r'^2}\\ &= r\sqrt{1 - \frac{2\hat{r}\cdot r'}{r} + \frac{r'^2}{r^2}}\\ &\sim r\left[1 - \frac{\hat{r}\cdot r'}{r} + O\left(\frac{1}{r^2}\right)\right] .\end{aligned} \tag{3.15}$$

Note that in the last equation,

$$\hat{r} = \frac{r}{r} . \tag{3.16}$$

Let us introduce $\mathcal{K}$, the wave number in the direction of the outgoing unit vector $\hat{r}$,

$$\mathcal{K} \equiv K\hat{r} . \tag{3.17}$$

So the Green function for the operator $\nabla^2 + K^2$, expressed by (3.9), has the following asymptotic behaviour:

$$g(r,r') \underset{r\to\infty}{\sim} -\frac{\exp(iKr - i\mathcal{K}\cdot r')}{4\pi r} . \tag{3.18}$$

Let us now introduce the asymptotic behaviour of the Green function (3.18) into the integral equation (3.8):

$$\Psi(r) \underset{r\to\infty}{\sim} \exp(iKz) - \frac{2m}{\hbar^2}\int d^3r' \frac{\exp(iKr - i\mathcal{K}\cdot r')}{4\pi r}V(r')\Psi(r') . \tag{3.19}$$

From the equation

$$\begin{aligned}&\int d^3r' \frac{\exp(iKr - i\mathcal{K}\cdot r')}{4\pi r}V(r')\Psi(r')\\ &= \frac{\exp(iKr)}{r}\int d^3r' \frac{\exp(-i\mathcal{K}\cdot r')}{4\pi}V(r')\Psi(r') ,\end{aligned} \tag{3.20}$$

we can conclude that, if the scattering amplitude is given by

$$f(\theta, \phi) = -\frac{m}{2\pi\hbar^2} \int d^3r \exp(-i\boldsymbol{\mathcal{K}} \cdot \boldsymbol{r}) V(\boldsymbol{r}) \Psi(\boldsymbol{r}) , \qquad (3.21)$$

then the boundary conditions are satisfied. In (3.21), $\boldsymbol{\mathcal{K}}$ is the wave number of the scattered particles, and $\Psi(\boldsymbol{r})$ is the scattering wave function.

Let us suppose that the ratio between the electron kinetic energy and the atomic potential energy is high enough to render the scattering weak and $\Psi(\boldsymbol{r})$ not very different from the incident plane wave $\exp(iKz)$. This is the assumption which is the basis of the first Born approximation, i.e.

$$\Psi(\boldsymbol{r}) = \exp(iKz) = \exp(i\boldsymbol{K} \cdot \boldsymbol{r}) . \qquad (3.22)$$

Utilising the first Born approximation, expressed by (3.22), the previous equation (3.21) becomes

$$f(\theta, \phi) = -\frac{m}{2\pi\hbar^2} \int d^3r \exp(-i\boldsymbol{\mathcal{K}} \cdot \boldsymbol{r}) V(\boldsymbol{r}) \exp(i\boldsymbol{K} \cdot \boldsymbol{r}) . \qquad (3.23)$$

If we use $\hbar\boldsymbol{q}$ to indicate the momentum lost by the incident electron,

$$\hbar\boldsymbol{q} = \hbar(\boldsymbol{K} - \boldsymbol{\mathcal{K}}) , \qquad (3.24)$$

for fast particles we can write that

$$f(\theta, \phi) = -\frac{m}{2\pi\hbar^2} \int d^3r \exp(i\boldsymbol{q} \cdot \boldsymbol{r}) V(\boldsymbol{r}) . \qquad (3.25)$$

As we are interested in a central potential, then

$$V(\boldsymbol{r}) = V(r) , \qquad (3.26)$$

and as a result,

$$f(\theta, \phi) = f(\theta)$$
$$= -\frac{m}{2\pi\hbar^2} \int_0^{2\pi} d\phi \int_0^{\pi} \sin\theta \, d\theta \int_0^{\infty} r^2 \, dr \, \exp(iqr\cos\theta) V(r) . \qquad (3.27)$$

We carry out the integrations over ϕ and over θ and obtain

$$f(\theta) = -\frac{2m}{\hbar^2 q} \int_0^{\infty} \sin(qr) V(r) \, r \, dr . \qquad (3.28)$$

We are interested in the calculation of differential elastic scattering cross-sections in the first Born approximation for a screened Coulomb potential, such as a Wentzel-like potential [1],

$$V(r) = -\frac{Ze^2}{r} \exp\left(-\frac{r}{a}\right) . \qquad (3.29)$$

The exponential factor here represents a rough approximation of the screening of the nucleus by the orbital electrons, while the a parameter is

$$a = \frac{a_0}{Z^{1/3}} , \qquad (3.30)$$

where $a_0 = \hbar^2/me^2$ is the Bohr radius.

3.1 The First Born Approximation

Let us calculate the scattering amplitude:

$$f(\theta) = \frac{2m}{\hbar^2} \frac{Ze^2}{q} \int_0^\infty \sin(qr) \exp\left(-\frac{r}{a}\right) dr . \tag{3.31}$$

As the equation

$$\int_0^\infty \sin(qr) \exp\left(-\frac{r}{a}\right) dr = \frac{q}{q^2 + (1/a)^2} \tag{3.32}$$

holds, we can conclude that

$$\frac{d\sigma}{d\Omega} = |f(\theta)|^2 = \frac{4m^2}{\hbar^4} \frac{Z^2 e^4}{[q^2 + (1/a)^2]^2} . \tag{3.33}$$

On the other hand, $|\boldsymbol{K}| = |\boldsymbol{\mathcal{K}}|$ and $\boldsymbol{q} = \boldsymbol{K} - \boldsymbol{\mathcal{K}}$, and, as a consequence

$$\begin{aligned}
q^2 &= (\boldsymbol{K} - \boldsymbol{\mathcal{K}}) \cdot (\boldsymbol{K} - \boldsymbol{\mathcal{K}}) \\
&= \boldsymbol{K}^2 + \boldsymbol{\mathcal{K}}^2 - 2K\mathcal{K}\cos\theta \\
&= 2\boldsymbol{K}^2 (1 - \cos\theta) ,
\end{aligned} \tag{3.34}$$

where θ is the scattering angle.

The electron kinetic energy is given by

$$E = \frac{\hbar^2 \boldsymbol{K}^2}{2m} , \tag{3.35}$$

so that the differential elastic scattering cross-section for the collision of an electron beam with a Wentzel-like atomic potential is given in the first Born approximation by

$$\frac{d\sigma}{d\Omega} = \frac{Z^2 e^4}{4E^2} \frac{1}{(1 - \cos\theta + \alpha)^2} . \tag{3.36}$$

In the last equation, α is the *screening parameter*, given by

$$\alpha = \frac{1}{2K^2 a^2} = \frac{me^4 \pi^2}{h^2} \frac{Z^{2/3}}{E} . \tag{3.37}$$

The well-known classical Rutherford formula,

$$\frac{d\sigma}{d\Omega} = \frac{Z^2 e^4}{4E^2} \frac{1}{(1 - \cos\theta)^2} , \tag{3.38}$$

can be obtained by imposing $\alpha = 0$ in (3.36).

The total elastic scattering cross-section can be obtained from

$$\sigma_{el} = \int \frac{d\sigma}{d\Omega} d\Omega . \tag{3.39}$$

For a Wentzel-like potential, the total elastic scattering cross-section can be easily calculated thus:

$$\sigma_{el} = \frac{Z^2 e^4}{4E^2} \int_0^{2\pi} d\phi \int_0^\pi \sin\vartheta \, d\vartheta \frac{1}{(1-\cos\vartheta+\alpha)^2}$$
$$= \frac{\pi Z^2 e^4}{E^2} \frac{1}{\alpha(2+\alpha)} \ . \tag{3.40}$$

If $\alpha \to 0$ and, as a consequence, the differential elastic scattering cross-section is given by the classical Rutherford formula, the total elastic scattering cross-section diverges, reflecting the long-range nature of the pure Coulomb potential.

The elastic mean free path is the reciprocal of the total elastic scattering cross-section divided by the number N of atoms per unit of volume in the target:

$$\lambda_{el} = \frac{1}{N\sigma_{el}} = \frac{\alpha(2+\alpha)E^2}{N\pi e^4 Z^2} \ . \tag{3.41}$$

3.2 The Density Matrix and Spin Polarisation

Following the previous section concerning the scattering of a spinless particle, it is necessary to introduce the density matrix, the spin polarisation and relativistic phenomena in order to take into account the effect of spin on the elastic scattering.

The spin orientation of an electron beam is typically known through a probability distribution, which means it cannot be specified by a single state vector. Instead, it is a quantum system in a mixed state, constituted by n subsystems, each of which is in a pure state.

Let us indicate by $|a\rangle$ the normalised state vectors of the systems in pure states. If we denote by $|i\rangle$ a complete set of orthonormal eigenvectors, then we can expand each pure state as follows:

$$|a\rangle = \sum_i c_i |i\rangle \ , \tag{3.42}$$

where the coefficients of the expansion c_i are the projections of $|a\rangle$ on $|i\rangle$:

$$c_i = \langle i|a\rangle \ . \tag{3.43}$$

Let us now consider an operator A and carry out an evaluation of its average value

$$\langle A \rangle = \sum_{a=1}^n p_a \langle a|A|a\rangle \ , \tag{3.44}$$

in which we have denoted by p_a the probability of obtaining the pure state $|a\rangle$.

3.2 The Density Matrix and Spin Polarisation

Let us now define the density operator

$$\mu = \sum_{a=1}^{n} |a\rangle p_a \langle a| . \tag{3.45}$$

Before proceeding, we can observe that the matrix elements of the density operator constitute the density matrix and are given by

$$\mu_{ii'} = \langle i|\mu|i'\rangle = \sum_{a=1}^{n} \langle i|a\rangle p_a \langle a|i'\rangle = \sum_{a=1}^{n} p_a c_i'^* c_i . \tag{3.46}$$

It is possible to see that the average value of operator A can be calculated as the trace of operator μA, once the density operator is introduced. The set of orthonormal eigenvectors denoted by $|i\rangle$ is supposed to be complete. As a consequence,

$$\sum_i |i\rangle\langle i| = 1 . \tag{3.47}$$

So, the average value of operator A is given by

$$\begin{aligned}
\langle A \rangle &= \sum_{a=1}^{n} \sum_i \sum_{i'} \langle i|a\rangle p_a \langle a|i'\rangle \langle i'|A|i\rangle \\
&= \sum_i \sum_{i'} \langle i|\mu|i'\rangle \langle i'|A|i\rangle \\
&= \sum_i \langle i|\mu A|i\rangle \\
&= \mathrm{Tr}(\mu A) .
\end{aligned} \tag{3.48}$$

If we now consider the spin space, a complete set of 2×2 Hermitian matrices is given by the Pauli matrices σ_i and the unit matrix I, so that the density matrix may be written as

$$\mu = v_0 I + \sum_i v_i \sigma_i , \tag{3.49}$$

where the v_i, with $i = 0, 1, 2, 3$, are real coefficients.

The average values of the Pauli matrices are given by

$$\langle \sigma_i \rangle = \mathrm{Tr}(\mu \sigma_i) . \tag{3.50}$$

The value of v_0 can be obtained by observing that

$$\mathrm{Tr}(\mu) = \mathrm{Tr}(\mu I) = \langle I \rangle = 1 , \tag{3.51}$$

and

$$\mathrm{Tr}(\sigma_i) = 0 . \tag{3.52}$$

As a consequence,

$$1 = \mathrm{Tr}(\mu) = v_0 \mathrm{Tr}(I) = 2v_0 , \tag{3.53}$$

or
$$v_0 = \frac{1}{2}. \tag{3.54}$$

Concerning the calculation of the values of the parameters v_i with $i = 1, 2, 3$, we can observe that
$$\text{Tr}(\sigma_i \sigma_j) = 2\delta_{ij}. \tag{3.55}$$

Consequently
$$\langle \sigma_i \rangle = \text{Tr}(\mu \sigma_i) = 2v_i, \tag{3.56}$$

or
$$v_i = \frac{1}{2} \langle \sigma_i \rangle. \tag{3.57}$$

If we now introduce the components of the polarisation vector, defined as
$$P_i \equiv \langle \sigma_i \rangle = \text{Tr}(\mu \sigma_i), \tag{3.58}$$

the density matrix may be written as
$$\mu = \frac{1}{2}\left(I + \sum_i \sigma_i P_i\right), \tag{3.59}$$

or
$$\mu = \frac{1}{2}\begin{pmatrix} 1 + P_3 & P_1 - iP_2 \\ P_1 + iP_2 & 1 - P_3 \end{pmatrix}. \tag{3.60}$$

Choosing the z axis in the direction of the polarisation vector $\boldsymbol{P}$, we obtain $P_1 = P_2 = 0$, $P_3 = |\boldsymbol{P}| = P$ and
$$\mu = \frac{1}{2}\begin{pmatrix} 1 + P & 0 \\ 0 & 1 - P \end{pmatrix}. \tag{3.61}$$

The diagonal elements of the density matrix,
$$\mu_{ii} = \langle i | \mu | i \rangle, \tag{3.62}$$

represent the probability that an electron of the beam is found in state i. As a consequence, if u indicates the number of electrons of the beam with spin up, and d the number of electrons of the beam with spin down, we can write
$$\frac{1+P}{2} = \frac{u}{u+d}, \tag{3.63}$$

so that the polarisation P of the beam is given by
$$P = \frac{u - d}{u + d}. \tag{3.64}$$

Note that μ can be decomposed as follows:
$$\mu = (1 - P)\mu_{1/2} + P\mu_1 = (1 - P)\begin{pmatrix} 1/2 & 0 \\ 0 & 1/2 \end{pmatrix} + P\begin{pmatrix} 1 & 0 \\ 0 & 0 \end{pmatrix}. \tag{3.65}$$

The diagonal elements of the matrix $\mu_{1/2}$,

$$\mu_{1/2} = \begin{pmatrix} 1/2 & 0 \\ 0 & 1/2 \end{pmatrix}, \qquad (3.66)$$

correspond to a completely unpolarised system, since the probabilities of spin up and spin down are in this case equal to $1/2$. The diagonal elements of the second matrix μ_1,

$$\mu_1 = \begin{pmatrix} 1 & 0 \\ 0 & 0 \end{pmatrix}, \qquad (3.67)$$

correspond to a pure state in which all the electrons of the beam have spin up; i.e. for a density matrix equal to μ_1, the spin projection is in the direction of $\boldsymbol{P}$ for all the electrons of the beam. The beam in this case is said to be totally polarised. When $P = 0$ and the density matrix is $\mu_{1/2}$, the beam is completely unpolarised. When $P = 1$ and the density matrix is μ_1, or $P = -1$, the beam is totally polarised. In all other situations, corresponding to $0 < |P| < 1$, the beam is said to be partially polarised.

3.3 Relativistic Partial-Wave Expansion

The fundamental equation of relativistic quantum mechanics is the Dirac equation. The wave function, as is well known, is a four-component spinor, and the asymptotic forms of the four components of the scattered wave are

$$\Psi_i \underset{r \to \infty}{\sim} a_i \exp(iKz) + b_i(\theta, \phi) \frac{\exp(iKr)}{r}. \qquad (3.68)$$

The differential elastic scattering cross-section is given by

$$\begin{aligned}
\frac{d\sigma}{d\Omega} &= \frac{|b_1|^2 + |b_2|^2 + |b_3|^2 + |b_4|^2}{|a_1|^2 + |a_2|^2 + |a_3|^2 + |a_4|^2} \\
&= \frac{|b_1|^2 + |b_2|^2 + c|b_1|^2 + c|b_2|^2}{|a_1|^2 + |a_2|^2 + c|a_1|^2 + c|a_2|^2} \\
&= \frac{|b_1|^2 + |b_2|^2}{|a_1|^2 + |a_2|^2},
\end{aligned} \qquad (3.69)$$

where c is a constant of proportionality which takes into account the fact that the a_i and the b_i coefficients are not all independent. Indeed, asymptotically the scattered wave is made up of plane waves proceeding, from the centre, in various directions; and the coefficients of the solutions for a plane wave are not all independent.

If the spin is parallel to the direction of incidence (spin up), $a_1 = 1$, $a_2 = 0$, $b_1 = f^+(\theta, \phi)$, $b_2 = g^+(\theta, \phi)$, where f^+ and g^+ are two scattering amplitudes.

The asymptotic behaviour is described by the following equations:

$$\Psi_1 \underset{r\to\infty}{\sim} \exp(iKz) + f^+(\theta,\phi)\frac{\exp(iKr)}{r}, \qquad (3.70)$$

$$\Psi_2 \underset{r\to\infty}{\sim} g^+(\theta,\phi)\frac{\exp(iKr)}{r}. \qquad (3.71)$$

The case of spin antiparallel to the direction of incidence (spin down) corresponds to $a_1 = 0$, $a_2 = 1$, $b_1 = g^-(\theta,\phi)$, $b_2 = f^-(\theta,\phi)$. The asymptotic behaviour is now given by

$$\Psi_1 \underset{r\to\infty}{\sim} g^-(\theta,\phi)\frac{\exp(iKr)}{r}, \qquad (3.72)$$

$$\Psi_2 \underset{r\to\infty}{\sim} \exp(iKz) + f^-(\theta,\phi)\frac{\exp(iKr)}{r}. \qquad (3.73)$$

The Dirac equations for an electron in a central field are given by the following (see Chap. 2):

$$[E + m - V(r)]F_l^\pm(r) + \frac{dG_l^\pm(r)}{dr} + \frac{1+k}{r}G_l^\pm(r) = 0, \qquad (3.74)$$

$$-[E - m - V(r)]G_l^\pm(r) + \frac{dF_l^\pm(r)}{dr} + \frac{1-k}{r}F_l^\pm(r) = 0. \qquad (3.75)$$

The natural units $\hbar = c = 1$ are used here, as they are particularly convenient for the quantum-relativistic equations. The superscript "+" refers to the electrons with spin up $(k = -(l+1))$ while "−" refers to electrons with spin down $(k = l)$. Once the new variables

$$\mu(r) \equiv E + m - V(r), \qquad (3.76)$$

$$\nu(r) \equiv E - m - V(r), \qquad (3.77)$$

$$\mu' = \frac{d\mu}{dr} \qquad (3.78)$$

have been introduced, the Dirac equations become

$$F_l^\pm(r) = -\frac{1}{\mu}\left(\frac{dG_l^\pm}{dr} + \frac{1+k}{r}G_l^\pm\right) \qquad (3.79)$$

and

$$\frac{dF_l^\pm}{dr} = \frac{\mu'}{\mu^2}\left(\frac{dG_l^\pm}{dr} + \frac{1+k}{r}G_l^\pm\right)$$

$$-\frac{1}{\mu}\left(\frac{d^2G_l^\pm}{dr^2} + \frac{1+k}{r}\frac{dG_l^\pm}{dr} - \frac{1+k}{r^2}G_l^\pm\right). \qquad (3.80)$$

3.3 Relativistic Partial-Wave Expansion

Therefore, after simple algebraic manipulations, we obtain the following:

$$\frac{d^2 G_l^\pm}{dr^2} + \left(\frac{2}{r} - \frac{\mu'}{\mu}\right)\frac{dG_l^\pm}{dr} + \left(\mu\nu - \frac{k(k+1)}{r^2} - \frac{1+k}{r}\frac{\mu'}{\mu}\right) G_l^\pm = 0 . \quad (3.81)$$

Let us now introduce the function $\mathcal{G}_l^\pm$, where

$$\mathcal{G}_l^\pm \equiv \frac{r}{\mu^{1/2}} G_l^\pm . \quad (3.82)$$

Upon observing that

$$K^2 = E^2 - m^2 , \quad (3.83)$$

it is possible to see that

$$\mu\nu = K^2 - 2EV + V^2 . \quad (3.84)$$

We conclude that, once the function $U_l^\pm(r)$ has been defined, i.e.

$$-U_l^\pm(r) = -2EV + V^2 - \frac{k}{r}\frac{\mu'}{\mu} + \frac{1}{2}\frac{\mu''}{\mu} - \frac{3}{4}\frac{\mu'^2}{\mu^2} , \quad (3.85)$$

the following equation holds:

$$\left[\frac{d^2}{dr^2} - \frac{k(k+1)}{r^2} + K^2 - U_l^\pm(r)\right] \mathcal{G}_l^\pm = 0 . \quad (3.86)$$

For large values of r, $\mathcal{G}_l^\pm$ is essentially sinusoidal. Indeed, when r is large enough, $V(r)$ is negligible, $U_l^\pm$ is almost constant and the solution of the equation is therefore a linear combination of the regular and irregular spherical Bessel functions multiplied by Kr (see Appendix C). Taking account of the fact that $\mathcal{G}_l^\pm = (r/\mu^{1/2}) G_l^\pm$, we can therefore conclude that

$$G_l^\pm \underset{r\to\infty}{\sim} j_l(Kr)\cos\eta_l^\pm - n_l(Kr)\sin\eta_l^\pm . \quad (3.87)$$

Here $\eta_l^\pm$ are constants to be determined. Taking into account the asymptotic behaviour of the Bessel functions,

$$j_l(Kr) \underset{r\to\infty}{\sim} \frac{1}{Kr}\sin\left(Kr - \frac{l\pi}{2}\right) , \quad (3.88)$$

$$n_l(Kr) \underset{r\to\infty}{\sim} -\frac{1}{Kr}\cos\left(Kr - \frac{l\pi}{2}\right) , \quad (3.89)$$

we can conclude that

$$G_l^\pm \underset{r\to\infty}{\sim} \frac{1}{Kr}\sin\left(Kr - \frac{l\pi}{2}\right)\cos\eta_l^\pm + \frac{1}{Kr}\cos\left(Kr - \frac{l\pi}{2}\right)\sin\eta_l^\pm . \quad (3.90)$$

Therefore,
$$G_l^+ \underset{r\to\infty}{\sim} \frac{1}{Kr} \sin\left(Kr - \frac{l\pi}{2} + \eta_l^+\right), \qquad (3.91)$$

and
$$G_l^- \underset{r\to\infty}{\sim} \frac{1}{Kr} \sin\left(Kr - \frac{l\pi}{2} + \eta_l^-\right). \qquad (3.92)$$

The phase shifts $\eta_l^\pm$ represent the effect of the potential $V(r)$ on the phases of the scattered waves.

Before proceeding, we need to demonstrate the following equation:
$$\exp(iKr\cos\theta) = \sum_{l=0}^{\infty}(2l+1)i^l j_l(Kr)P_l(\cos\theta), \qquad (3.93)$$

where $P_l(\cos\theta)$ are the Legendre polynomials (see Appendix C) and $j_l(Kr)$ the spherical Bessel functions. In order to demonstrate this, let us first observe that a plane wave describing a free particle with the z axis in the direction of $\boldsymbol{K}$ may be expressed as an expansion in a series of Legendre polynomials $P_l(\cos\theta)$:
$$\exp(iKz) = \exp(iKr\cos\theta) = \sum_{l=0}^{\infty} c_l j_l(Kr)P_l(\cos\theta). \qquad (3.94)$$

Let us define the two new variables $s \equiv Kr$ and $t \equiv \cos\theta$, to have
$$\exp(ist) = \sum_l c_l j_l(s)P_l(t). \qquad (3.95)$$

Differentiating with respect to s,
$$it\exp(ist) = \sum_l it c_l j_l(s)P_l(t) = \sum_l c_l \frac{dj_l(s)}{ds}P_l(t). \qquad (3.96)$$

Recalling one of the properties of the Legendre polynomials (see Appendix C), we have:
$$P_l(t) = \frac{(l+1)P_{l+1}(t) + lP_{l-1}(t)}{t(2l+1)}. \qquad (3.97)$$

Therefore,
$$\begin{aligned}it\exp(ist) &= \sum_l it c_l j_l(s) \frac{(l+1)P_{l+1}(t) + lP_{l-1}(t)}{t(2l+1)}\\ &= \sum_l i P_l(t)\left[\frac{l}{2l-1}c_{l-1}j_{l-1}(s) + \frac{l+1}{2l+3}c_{l+1}j_{l+1}(s)\right]. \end{aligned} \qquad (3.98)$$

3.3 Relativistic Partial-Wave Expansion

On the other hand, it is well known that

$$\frac{dj_l(s)}{ds} = \frac{l}{2l+1} j_{l-1}(s) - \frac{l+1}{2l+1} j_{l+1}(s) , \qquad (3.99)$$

and as a consequence,

$$it \exp(ist) = \sum_l c_l P_l(t) \left[\frac{l}{2l+1} j_{l-1}(s) - \frac{l+1}{2l+1} j_{l+1}(s) \right] . \qquad (3.100)$$

Consequently, from (3.98) and (3.100), the following is obtained:

$$\sum_l P_l(t) \left[j_{l-1}(s) l \left(\frac{c_l}{2l+1} - \frac{ic_{l-1}}{2l-1} \right) \right.$$
$$\left. - j_{l+1}(s)(l+1) \left(\frac{c_l}{2l+1} + \frac{ic_{l+1}}{2l+3} \right) \right] = 0 . \qquad (3.101)$$

The Legendre polynomials $P_l(t)$ are linearly independent (orthonormal), and therefore

$$j_{l-1}(s) l \left(\frac{c_l}{2l+1} - \frac{ic_{l-1}}{2l-1} \right)$$
$$= j_{l+1}(s) (l+1) \left(\frac{c_l}{2l+1} + \frac{ic_{l+1}}{2l+3} \right) . \qquad (3.102)$$

Every value of s satisfies the last equation if and only if

$$\frac{1}{2l+1} c_l = \frac{i}{2l-1} c_{l-1} . \qquad (3.103)$$

In order to obtain an explicit expression for c_l we need to know the value of the first coefficient of the set, i.e. c_0. Imposing $r = 0$ in (3.94), we obtain

$$\exp(0) = 1 = \sum_l c_l j_l(0) P_l(\cos\theta) . \qquad (3.104)$$

Since $j_l(0) = 0$ for any $l \neq 0$, while $j_0(0) = 1$ and $P_0(\cos\theta) = 1$, we may conclude that $c_0 = 1$. The recursive repetition of the relation (3.103) allows us to obtain the values of the coefficients,

$$c_l = (2l+1) i^l , \qquad (3.105)$$

and the expansion of the plane wave in Legendre polynomials,

$$\exp(iKr\cos\theta) = \exp(iKz) = \sum_{l=0}^{\infty} (2l+1) i^l j_l(Kr) P_l(\cos\theta) . \qquad (3.106)$$

Let us remind the reader that we are looking for the functions Ψ_1 and Ψ_2 which satisfy the asymptotic conditions. So, we must begin by expanding them in spherical harmonics (see Appendix C):

$$\Psi_1 = \sum_{l=0}^{\infty} [A_l G_l^+ + B_l G_l^-] P_l(\cos\theta) , \tag{3.107}$$

$$\Psi_2 = \sum_{l=1}^{\infty} [C_l G_l^+ + D_l G_l^-] P_l^1(\cos\theta) \exp(i\phi) . \tag{3.108}$$

The coefficients A_l, B_l, C_l and D_l can be determined by considering the asymptotic behaviours of the functions involved. Let us begin with the function Ψ_1 and observe that

$$\Psi_1 - \exp(iKz) = \sum_{l=0}^{\infty} [A_l G_l^+ + B_l G_l^- - (2l+1)i^l j_l(Kr)] P_l(\cos\theta) . \tag{3.109}$$

As

$$\Psi_1 - \exp(iKz) \underset{r \to \infty}{\sim} \frac{\exp(iKr)}{r} f^+(\theta, \phi) , \tag{3.110}$$

the following occurs:

$$\frac{1}{Kr} \sum_{l=0}^{\infty} \left[A_l \sin\left(Kr - \frac{l\pi}{2} + \eta_l^+\right) + B_l \sin\left(Kr - \frac{l\pi}{2} + \eta_l^-\right) \right.$$
$$\left. -(2l+1)i^l \sin\left(Kr - \frac{l\pi}{2}\right) \right] P_l(\cos\theta)$$
$$= \frac{\exp(iKr)}{r} f^+(\theta, \phi) . \tag{3.111}$$

As a consequence,

$$\frac{\exp(iKr)}{2iKr} \sum_{l=0}^{\infty} \exp\left(-i\frac{l\pi}{2}\right)$$
$$\times [A_l \exp(i\eta_l^+) + B_l \exp(i\eta_l^-) - (2l+1)i^l] P_l(\cos\theta)$$
$$- \frac{\exp(-iKr)}{2iKr} \sum_{l=0}^{\infty} \exp\left(i\frac{l\pi}{2}\right)$$
$$\times [A_l \exp(-i\eta_l^+) + B_l \exp(-i\eta_l^-) - (2l+1)i^l] P_l(\cos\theta)$$
$$= \frac{\exp(iKr)}{r} f^+(\theta, \phi) . \tag{3.112}$$

The asymptotic conditions are satisfied if

$$A_l \exp(-i\eta_l^+) + B_l \exp(-i\eta_l^-) = (2l+1)i^l . \tag{3.113}$$

3.3 Relativistic Partial-Wave Expansion

With the choices

$$A_l = li^l \exp(i\eta_l^+),\tag{3.114}$$

$$B_l = (l+1)i^l \exp(i\eta_l^-),\tag{3.115}$$

(3.113) is satisfied.

Proceeding in a similar way regarding the Ψ_2 function, we can therefore choose

$$C_l = i^l \exp(i\eta_l^+),\tag{3.116}$$

$$D_l = -i^l \exp(i\eta_l^-).\tag{3.117}$$

In conclusion, for electrons with spins parallel to the direction of incidence, we have

$$\Psi_1 = \sum_{l=0}^{\infty}[(l+1)\exp(i\eta_l^-)G_l^- + l\exp(i\eta_l^+)G_l^+]i^l P_l(\cos\theta),\tag{3.118}$$

$$\Psi_2 = \sum_{l=1}^{\infty}[\exp(i\eta_l^+)G_l^+ - \exp(i\eta_l^-)G_l^-]i^l P_l^1(\cos\theta)\exp(i\phi),\tag{3.119}$$

and, by using (3.112),

$$f^+(\theta,\phi) = f^+(\theta)$$
$$= \frac{1}{2iK}\sum_{l=0}^{\infty}\{(l+1)[\exp(2i\eta_l^-)-1]$$
$$+l[\exp(2i\eta_l^+)-1]\}P_l(\cos\theta),\tag{3.120}$$

$$g^+(\theta,\phi) = \frac{1}{2iK}\sum_{l=1}^{\infty}[\exp(2i\eta_l^+) - \exp(2i\eta_l^-)]P_l^1(\cos\theta)\exp(i\phi).\tag{3.121}$$

For electrons with spins antiparallel to the direction of incidence (spin down), where we indicate the scattering amplitudes by f^- and g^-, it is possible to see that

$$f^-(\theta,\phi) = f^+(\theta,\phi)\tag{3.122}$$

and

$$g^-(\theta,\phi) = -g^+(\theta,\phi)\exp(-2i\phi).\tag{3.123}$$

It is therefore convenient to define the functions

$$f(\theta) = \sum_{l=0}^{\infty} A_l P_l(\cos\theta),\tag{3.124}$$

$$g(\theta) = \sum_{l=0}^{\infty} B_l P_l^1(\cos\theta),\tag{3.125}$$

where

$$\mathcal{A}_l = \frac{1}{2iK}\{(l+1)[\exp(2i\eta_l^-) - 1] + l[\exp(2i\eta_l^+) - 1]\}, \quad (3.126)$$

$$\mathcal{B}_l = \frac{1}{2iK}\{\exp(2i\eta_l^+) - \exp(2i\eta_l^-)\}. \quad (3.127)$$

With this notation, we have

$$f^+ = f^- = f, \quad (3.128)$$

$$g^+ = g\exp(i\phi) \quad (3.129)$$

and

$$g^- = -g\exp(-i\phi). \quad (3.130)$$

For an arbitrary spin direction, the electron incident plane wave will be given by $\Psi_1 = A\exp(iKz)$ and $\Psi_2 = B\exp(iKz)$, and as a consequence $a_1 = A$, $a_2 = B$. Furthermore,

$$b_1 = Af^+ + Bg^- = Af - Bg\exp(-i\phi), \quad (3.131)$$

$$b_2 = Ag^+ + Bf^- = Bf + Ag\exp(i\phi). \quad (3.132)$$

Consequently,

$$\frac{d\sigma}{d\Omega} = (|f|^2 + |g|^2)\left\{1 + iS(\theta)\left[\frac{AB^*\exp(i\phi) - A^*B\exp(-i\phi)}{|A|^2 + |B|^2}\right]\right\}, \quad (3.133)$$

where $S(\theta)$ is the Sherman function, defined by

$$S(\theta) = i\frac{fg^* - f^*g}{|f|^2 + |g|^2}. \quad (3.134)$$

Note that

$$i\frac{AB^*\exp(i\phi) - A^*B\exp(-i\phi)}{|A|^2 + |B|^2} = \xi^\dagger(\sigma_2\cos\phi - \sigma_1\sin\phi)\xi, \quad (3.135)$$

where σ_1, σ_2 and σ_3 are the Pauli matrices and ξ is the two-component spinor

$$\xi = \begin{pmatrix} A/\sqrt{|A|^2 + |B|^2} \\ B/\sqrt{|A|^2 + |B|^2} \end{pmatrix}, \quad (3.136)$$

$$\xi^\dagger = \begin{pmatrix} \frac{A^*}{\sqrt{|A|^2 + |B|^2}} & \frac{B^*}{\sqrt{|A|^2 + |B|^2}} \end{pmatrix}. \quad (3.137)$$

As the z axis has been chosen along the incidence direction, the unit vector perpendicular to the plane of scattering is given by

$$\hat{n} = (-\sin\phi, \cos\phi, 0), \quad (3.138)$$

so we can therefore write

$$\xi^{\dagger}(\sigma_2 \cos\phi - \sigma_1 \sin\phi)\xi = \boldsymbol{P}\cdot\hat{\boldsymbol{n}}\,, \tag{3.139}$$

where $\boldsymbol{P}$ is the initial polarisation vector of the electron beam. The differential elastic scattering cross-section can then be recast in the following form:

$$\frac{d\sigma}{d\Omega} = (|f|^2 + |g|^2)[1 + S(\theta)\boldsymbol{P}\cdot\hat{\boldsymbol{n}}]\,. \tag{3.140}$$

Note that, if the beam is completely unpolarised, then $P = 0$ and

$$\frac{d\sigma}{d\Omega} = |f|^2 + |g|^2\,. \tag{3.141}$$

The total elastic scattering cross-section (σ_{el}) and the transport cross-section (σ_{tr}) are defined by

$$\sigma_{el} = 2\pi \int_0^\pi \frac{d\sigma}{d\Omega} \sin\theta\, d\theta\,, \tag{3.142}$$

$$\sigma_{tr} = 2\pi \int_0^\pi (1 - \cos\theta)\frac{d\sigma}{d\Omega} \sin\theta\, d\theta\,, \tag{3.143}$$

which can be easily calculated by numerical integration.

Note that by imposing

$$\eta_l^- = \eta_l^+ = \eta_l \tag{3.144}$$

in the previous equations, we can obtain the non-relativistic results. Indeed,

$$\mathcal{A}_l = \frac{1}{2iK}\{(l+1)[\exp(2i\eta_l) - 1] + l[\exp(2i\eta_l) - 1]\}$$

$$= \frac{1}{2iK}(2l+1)[\exp(2i\eta_l) - 1]\,, \tag{3.145}$$

$$\mathcal{B}_l = 0\,, \tag{3.146}$$

so that

$$f(\theta) = \frac{1}{2iK}\sum_{l=0}^{\infty}(2l+1)[\exp(2i\eta_l) - 1]P_l(\cos\theta)$$

$$= \frac{1}{K}\sum_{l=0}^{\infty}(2l+1)\exp(i\eta_l)\sin\eta_l P_l(\cos\theta)\,, \tag{3.147}$$

$$g(\theta) = 0 \tag{3.148}$$

and

$$\frac{d\sigma}{d\Omega} = |f|^2\,. \tag{3.149}$$

3.4 Calculation of the Phase Shifts

In order to proceed, it is convenient to perform the following transformation [2]:

$$F_l^\pm(r) = a_l^\pm(r) \frac{\sin \phi_l^\pm(r)}{r}, \tag{3.150}$$

$$G_l^\pm(r) = a_l^\pm(r) \frac{\cos \phi_l^\pm(r)}{r}. \tag{3.151}$$

After simple algebraic manipulations, (3.74) and (3.75) become

$$[E + m - V(r)] \tan \phi_l^\pm(r) + \frac{1}{a_l^\pm(r)} \frac{da_l^\pm(r)}{dr}$$
$$- \tan \phi_l^\pm(r) \frac{d\phi_l^\pm(r)}{dr} + \frac{k}{r} = 0, \tag{3.152}$$

$$-[E - m - V(r)] \cot \phi_l^\pm(r) + \frac{1}{a_l^\pm(r)} \frac{da_l^\pm(r)}{dr}$$
$$+ \cot \phi_l^\pm(r) \frac{d\phi_l^\pm(r)}{dr} - \frac{k}{r} = 0, \tag{3.153}$$

and therefore

$$\frac{d\phi_l^\pm(r)}{dr} = \frac{k}{r} \sin 2\phi_l^\pm(r) - m \cos 2\phi_l^\pm(r) + E - V(r), \tag{3.154}$$

$$\frac{1}{a_l^\pm(r)} \frac{da_l^\pm(r)}{dr} = -\frac{k}{r} \cos 2\phi_l^\pm(r) - m \sin 2\phi_l^\pm(r). \tag{3.155}$$

For $0 < r < \hbar/mc$, the spherical symmetric electrostatic potential experienced by a point charge at distance r from the nucleus, $V(r)$, may be approximated by the following:

$$V(r) \underset{r \to 0}{\sim} -\frac{Z_0 + Z_1 r + Z_2 r^2 + Z_3 r^3}{r}. \tag{3.156}$$

Expressing the electrostatic potential as the product of the potential of a bare nucleus multiplied by a screening function $\xi(r)$ having the analytical form

$$\xi(r) = \sum_{i=1}^{p} A_i \exp(-\alpha_i r), \tag{3.157}$$

$$\sum_{i=1}^{p} A_i = 1, \tag{3.158}$$

3.4 Calculation of the Phase Shifts

we can easily evaluate Z_0, Z_1, Z_2 and Z_3:

$$Z_0 = Ze^2 \sum_i A_i = Ze^2 , \tag{3.159}$$

$$Z_1 = -Z_0 \sum_{i=1}^{p} \alpha_i A_i , \tag{3.160}$$

$$Z_2 = \frac{Z_0}{2} \sum_{i=1}^{p} \alpha_i^2 A_i , \tag{3.161}$$

$$Z_3 = -\frac{Z_0}{6} \sum_{i=1}^{p} \alpha_i^3 A_i . \tag{3.162}$$

Let us expand $\phi_l^\pm$ as a power series

$$\phi_l^\pm(r) = \phi_{l0}^\pm + \phi_{l1}^\pm r + \phi_{l2}^\pm r^2 + \phi_{l3}^\pm r^3 + \ldots . \tag{3.163}$$

It is possible to see, after simple algebraic manipulations, that the relationships between the coefficients of this expansion and Z_0, Z_1, Z_2 and Z_3 are the following [3]:

$$\sin 2\phi_{l0}^\pm = -\frac{Z_0}{k} , \tag{3.164}$$

$$\phi_{l1}^\pm = \frac{E + Z_1 - m \cos 2\phi_{l0}^\pm}{1 - 2k \cos 2\phi_{l0}^\pm} , \tag{3.165}$$

$$\phi_{l2}^\pm = \frac{2\phi_{l1}^\pm \sin 2\phi_{l0}^\pm (m - k\phi_{l1}^\pm) + Z_2}{2 - 2k \cos 2\phi_{l0}^\pm} , \tag{3.166}$$

$$\phi_{l3}^\pm = \frac{2\phi_{l2}^\pm \sin 2\phi_{l0}^\pm (m - 2k\phi_{l1}^\pm) + 2\phi_{l1}^{\pm 2} \cos 2\phi_{l0}^\pm [m - (2/3)k\phi_{l1}^\pm] + Z_3}{3 - 2k \cos 2\phi_{l0}^\pm} , \tag{3.167}$$

with the extra conditions

$$0 \le 2\phi_{l0}^\pm \le \frac{1}{2}\pi \tag{3.168}$$

if $k < 0$, and

$$\pi \le 2\phi_{l0}^\pm \le \frac{3}{2}\pi \tag{3.169}$$

if $k > 0$.

Let us now calculate the phase shifts, examining (3.151):

$$G_l'^\pm = \frac{a_l'^\pm \cos \phi_l^\pm(r)}{r} - \frac{a_l^\pm}{r} \sin \phi_l^\pm(r) \phi_l'^\pm(r) - \frac{a_l^\pm \cos \phi_l^\pm(r)}{r^2} , \tag{3.170}$$

so that

$$\frac{G_l'^{\pm}}{G_l^{\pm}} = \frac{a_l'^{\pm}}{a_l^{\pm}} - \phi_l'^{\pm}(r)\tan\phi_l^{\pm}(r) - \frac{1}{r}, \qquad (3.171)$$

or

$$\frac{G_l'^{\pm}}{G_l^{\pm}} = -(E + m - V)\tan\phi_l^{\pm}(r) - \frac{1+k}{r}. \qquad (3.172)$$

Let us observe that the asymptotic form of the solution in the regions corresponding to large values of r for which $V(r) \approx 0$ is (see (3.87))

$$G_l^{\pm} = j_l(Kr)\cos\eta_l^{\pm} - n_l(Kr)\sin\eta_l^{\pm}, \qquad (3.173)$$

where $K^2 = E^2 - m^2$, $\eta_l^{\pm}$ are the lth phase shifts, and j_l and n_l are respectively the regular and irregular spherical Bessel functions. Therefore,

$$\frac{G_l'^{\pm}}{G_l^{\pm}} = \frac{K j_l'(Kr)\cos\eta_l^{\pm} - K n_l'(Kr)\sin\eta_l^{\pm}}{j_l(Kr)\cos\eta_l^{\pm} - n_l(Kr)\sin\eta_l^{\pm}}. \qquad (3.174)$$

Taking into account the properties of the Bessel functions

$$j_l'(x) = \frac{l}{x}j_l(x) - j_{l+1}(x), \qquad (3.175)$$

$$n_l'(x) = \frac{l}{x}n_l(x) - n_{l+1}(x), \qquad (3.176)$$

we may conclude that

$$\tan\eta_l^{\pm} = \frac{(l/r)j_l(Kr) - K j_{l+1}(Kr) - j_l(Kr)(G_l'^{\pm}/G_l^{\pm})}{(l/r)n_l(Kr) - K n_{l+1}(Kr) - n_l(Kr)(G_l'^{\pm}/G_l^{\pm})}. \qquad (3.177)$$

Let us define

$$\tilde{\phi}_l^{\pm} = \lim_{r \to \infty} \phi_l^{\pm}(r). \qquad (3.178)$$

For large values of r, (3.172) becomes

$$\frac{G_l'^{\pm}}{G_l^{\pm}} = -(E + m)\tan\tilde{\phi}_l^{\pm} - \frac{1+k}{r}, \qquad (3.179)$$

and therefore

$$\tan\eta_l^{\pm} = \frac{K j_{l+1}(Kr) - j_l(Kr)[(E+m)\tan\tilde{\phi}_l^{\pm} + (1+l+k)/r]}{K n_{l+1}(Kr) - n_l(Kr)[(E+m)\tan\tilde{\phi}_l^{\pm} + (1+l+k)/r]}. \qquad (3.180)$$

Using this last equation, we can calculate the phase shifts of the scattered wave and, therefore, the functions $f(\theta)$, $g(\theta)$ and the differential elastic scattering cross-section.

3.5 Exchange and Solid State Effects

The incident electron may be absorbed by the atom while another electron is emitted, so that such an exchange effect modifies the wave equations. This effect may be approximated, as proposed by Furness and McCarthy [4], by adding the following to the electron–atom potential energy:

$$V_{ex} = \frac{1}{2}(E - V_s) - \frac{1}{2}[(E - V_s)^2 + 4\pi\rho e^4 a_0]^{1/2} . \quad (3.181)$$

Here E represents the electron energy, V_s the electrostatic scalar potential, ρ the atomic electron density, e the electron charge and a_0 the Bohr radius.

Solid state effects should also be taken into account when the target atom is bound in a solid. An approximate way to treat this effect is to introduce what is known as central muffin-tin model, in which the potential of each atom of the solid is altered by the nearest neighbours; assuming that they are located at a distance $2r_{WS}$, the resulting potential is

$$V_{solid}(r) = V(r) + V(2r_{WS} - r) - 2V(r_{WS}) , \quad (3.182)$$

for $r \leq r_{WS}$, and zero elsewhere. Here r_{WS} is the Wigner–Seitz radius,

$$r_{WS} = 0.7346(A/\delta)^{1/3} \text{ Å} \quad (3.183)$$

where A is the atomic weight and δ the mass density expressed in g/cm^3. The term $2V(r_{WS})$ also has to be subtracted from the kinetic energy of the incident electron.

3.6 Comparing Theory and Experimental Data

In Figs. 3.1–3.5 we show a comparison of the electron differential elastic scattering cross-section calculated both from the screened Rutherford formula (first Born approximation) and by the relativistic partial-wave expansion method, with the experimental data available. The data concern different electron energies, target atoms and molecules. The energies considered are all below the Born threshold and it is evident from the comparison with experiment that the relativistic partial-wave expansion method allows a much more accurate description of the elastic scattering for these energies than does the screened Rutherford formula.

On the basis of this and other comparisons, and taking into account the experimental errors, we can say that the accuracy of the differential elastic scattering cross-sections is expected to be 1–2%, for scattering angles greater than 5°, as calculated by the relativistic partial-wave expansion method. The effect of the charge-induced atomic polarisation was neglected in our calculation code: such an effect may be relevant for angles lower than ∼5° and this means that we can expect an accuracy in the total cross-section of the order of 5–6%. In Fig. 3.6, we show a comparison of the total elastic scattering

38   3 Elastic Scattering

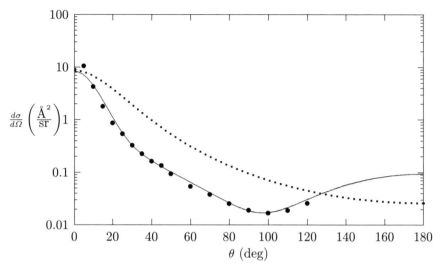

Fig. 3.1. Differential elastic scattering cross-section $d\sigma/d\Omega$ of 400 eV electrons scattered by Ar as a function of the scattering angle θ. *Dotted line*: screened Rutherford formula. *Solid line*: relativistic partial-wave expansion method [5]. •: Iga et al. experimental data [6]

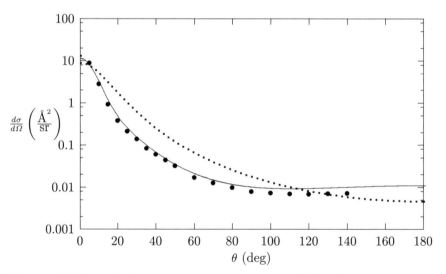

Fig. 3.2. Differential elastic scattering cross-section $d\sigma/d\Omega$ of 1000 eV electrons scattered by Ar as a function of the scattering angle θ. *Dotted line*: screened Rutherford formula. *Solid line*: relativistic partial-wave expansion method [5]. •: Iga et al. experimental data [6]

3.6 Comparing Theory and Experimental Data   39

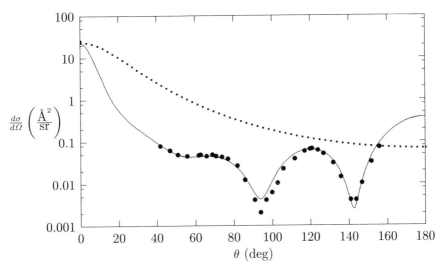

Fig. 3.3. Differential elastic scattering cross-section $d\sigma/d\Omega$ of 1100 eV electrons scattered by Au as a function of the scattering angle θ. *Dotted line*: screened Rutherford formula. *Solid line*: relativistic partial-wave expansion method [5]. •: Reichert experimental data [7]

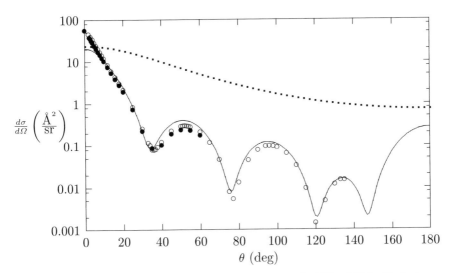

Fig. 3.4. Differential elastic scattering cross-section $d\sigma/d\Omega$ of 300 eV electrons scattered by Hg as a function of the scattering angle θ. *Dotted line*: screened Rutherford formula. *Solid line*: relativistic partial-wave expansion method [5]. •: Bromberg experimental data [8]. ○: Holtkamp et al. experimental data [9]

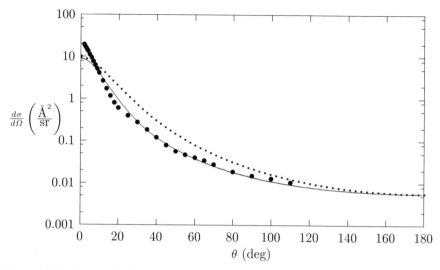

Fig. 3.5. Differential elastic scattering cross-section $d\sigma/d\Omega$ of 500 eV electrons scattered by CO as a function of the scattering angle θ. *Dotted line*: screened Rutherford formula. *Solid line*: relativistic partial-wave expansion method [5]. •: Bromberg experimental data [10]

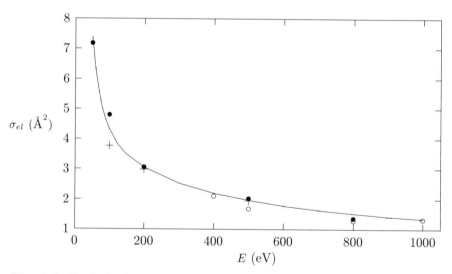

Fig. 3.6. Total elastic scattering cross-section σ_{el} of electrons scattered by Ar as a function of the electron kinetic energy E. *Solid line*: relativistic partial-wave expansion method [5]. +: Jansen et al. experimental data [11]. •: DuBois and Rudd experimental data [12]. ○: Iga et al. experimental data [6]

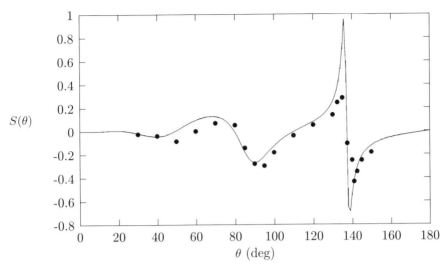

Fig. 3.7. Sherman function $S(\theta)$ for 1500 eV electrons scattered by gold as a function of the scattering angle θ. *Solid line*: relativistic partial-wave expansion method [5]. •: Deichsel and Reichert experimental data reported by Bunyan and Schonfelder [3]

cross-section calculated using the relativistic partial-wave expansion method and the experimental data of Jansen et al. [11], DuBois and Rudd [12] and Iga et al. [6].

The first transport cross-section is very important for multiple-scattering processes: this transport cross-section does not appear to be very much influenced by the inaccuracies of the differential elastic scattering cross-section at low scattering angle (owing to the factor $(1 - \cos\theta)$ in (3.143)), which are a consequence of neglecting atomic polarisation.

We are able to take into account the polarisation phenomena with the relativistic partial-wave expansion method. In Fig. 3.7, we compare the Sherman function (calculated with this method) with the Deichsel and Reichert experimental data reported by Bunyan and Schonfelder [3].

References

1. G. Wentzel, Z. Phys. **40**, 590 (1927)
2. S.-R. Lin, N. Sherman, J.K. Percus, Nucl. Phys. **45**, 492 (1963)
3. P.J. Bunyan, J.L. Schonfelder, Proc. Phys. Soc. **85**, 455 (1965)
4. J.B. Furness, I.E. McCarthy, J. Phys. B **6**, 2280 (1973)
5. M. Dapor, J. Appl. Phys. **79**, 8406 (1996)
6. I. Iga, Lee Mu-Tao, J.C. Nogueira, R.S. Barbieri, J. Phys. B **20**, 1095 (1987)
7. E. Reichert, Z. Phys. **173**, 392 (1963)
8. J.P. Bromberg, J. Chem. Phys. **51**, 4117 (1969)

9. G. Holtkamp, K. Jost, F.J. Peitzmann, J. Kessler, J. Phys. B: At. Mol. Phys. **20**, 4543 (1987)
10. J.P. Bromberg, J. Chem. Phys. **52**, 1243 (1970)
11. R.H.J. Jansen, F.J. de Heer, H.J. Luyken, B. van Wingerden, H.J. Blaauw, J. Phys. B: Atom. Mol. Phys. **9**, 185 (1976)
12. R.D. DuBois, M.E. Rudd, J. Phys. B: Atom. Mol. Phys. **9**, 2657 (1976)

4 Inelastic Scattering

4.1 The Classical Theory

Assume that a free electron is at rest. At a distance r from it, another electron is travelling along the z direction. If θ represents the angle between $\boldsymbol{r}$ and z, $\boldsymbol{v}$ represents the velocity of the incident electron and the instant of impact is $t = 0$,

$$-vt = r \cos \theta . \tag{4.1}$$

Defining the impact parameter b by

$$b \equiv r \sin \theta , \tag{4.2}$$

it is possible to see that b is the distance from the trajectory of the incident electron to the target:

$$b^2 + v^2 t^2 = r^2 . \tag{4.3}$$

$\boldsymbol{F}$ indicates the force of repulsion between the two electrons, where

$$F_z = \frac{e^2 \cos \theta}{r^2} \tag{4.4}$$

and

$$F_x = \frac{e^2 \sin \theta}{r^2} , \tag{4.5}$$

and p_z and p_x are the components of the momentum transferred to the target electron. As

$$\cot \theta = -\frac{vt}{b} , \tag{4.6}$$

differentiating with respect to θ, we can obtain

$$\frac{dt}{d\theta} = \frac{b}{v \sin^2 \theta} . \tag{4.7}$$

Therefore,

$$\frac{1}{r^2} \frac{dt}{d\theta} = \frac{1}{bv} . \tag{4.8}$$

The x component of $\boldsymbol{p}$ is thus given by

$$p_x = \int_{-\infty}^{\infty} F_x\, dt = \frac{e^2}{bv} \int_0^{\pi} \sin\theta\, d\theta = \frac{2e^2}{bv}, \qquad (4.9)$$

while the z component of $\boldsymbol{p}$ is zero. The energy W transferred to the electron at rest is

$$W = \frac{p_x^2}{2m} = \frac{e^4}{b^2 E}, \qquad (4.10)$$

where $E = mv^2/2$ is the kinetic energy of the incident electron.

The number of electrons in the volume $2\pi b\, db\, dz$ is given by $2\pi b\, db\, dz\, NZ$, where N is the number of atoms per unit of volume in the target and Z the target atomic number. Thus, the energy lost by the incident electron per unit of length in the target (called the *stopping power*) may be calculated from

$$-\frac{dE}{dz} = \int 2\pi b\, db\, W N Z = \frac{2\pi e^4 N Z}{E} \int_{b_{\min}}^{b_{\max}} \frac{db}{b}, \qquad (4.11)$$

where $b_{\min}$ and $b_{\max}$ are, respectively, the minimum and the maximum impact parameter.

The atomic electrons may be considered as a set of oscillators with frequencies ν_i and amplitudes f_i. As a consequence, we can rewrite the last equation as

$$-\frac{dE}{dz} = \frac{2\pi e^4 N}{E} \sum_i f_i \int_{b_{\min}}^{b_{\max}^i} \frac{db}{b}. \qquad (4.12)$$

The sum of the amplitudes of the oscillators has to be equal to the atomic number Z:

$$\sum_i f_i = Z. \qquad (4.13)$$

If $h\nu_i$ indicates the binding energy of the ith atomic electron, then

$$b_{\max}^i = v/\nu_i. \qquad (4.14)$$

The minimum impact parameter is given by

$$b_{\min} = h/mv, \qquad (4.15)$$

so that

$$-\frac{dE}{dz} = \frac{2\pi e^4 N}{E} \sum_i f_i \ln\left(\frac{mv^2}{h\nu_i}\right). \qquad (4.16)$$

Let us define the mean ionisation potential as

$$J^Z \equiv \prod_i (h\nu_i)^{f_i}, \qquad (4.17)$$

so that the stopping power becomes

$$-\frac{dE}{dz} = \frac{2\pi e^4 NZ}{E} \ln\left(\frac{2E}{J}\right). \tag{4.18}$$

This equation, deduced here with the classical theory, is substantially correct. Using a quantum mechanical treatment, the equation becomes the following:

$$-\frac{dE}{dz} = \frac{2\pi e^4 NZ}{E} \ln\left(\frac{CE}{J}\right), \tag{4.19}$$

where the constant C equals ≈ 1.166 (Bethe equation) [1]. The Bethe formula is valid for energies higher than $\sim J$, as it reaches a maximum for $E \approx 2.3J$, and then goes to zero for $E \approx J/1.166$: for lower energies the predicted stopping power becomes negative. Therefore, the low-energy stopping power requires a different approach, and its evaluation is based on the optical data of the material of interest.

4.2 Dielectric Function and Stopping Power

The response of a medium to an energy transfer $\hbar\omega$ and momentum transfer $\mathbf{q}$ is described by the complex dielectric function $\varepsilon(\mathbf{q}, \hbar\omega)$, which is related to the optical data. If $p(E, \hbar\omega)$ is the probability for an energy loss $\hbar\omega$ per unit distance travelled by an electron of energy E, the stopping power is then given by the following equation:

$$-\frac{dE}{ds} = \int \hbar\omega\, p(E, \hbar\omega)\, d\hbar\omega, \tag{4.20}$$

where the integration is over the allowed values of the energy transfer $\hbar\omega$. If $\hbar\omega$ is the energy transfer, q the momentum transfer and $\varepsilon(q, \hbar\omega)$ the complex dieletric function, then $p(E, \hbar\omega)$ is given by

$$p(E, \hbar\omega) = \frac{me^2}{\pi\hbar^2 E} \int_{q_1}^{q_2} \frac{dq}{q} \operatorname{Im}\left[\frac{-1}{\varepsilon(q, \hbar\omega)}\right], \tag{4.21}$$

where

$$q_1 = \sqrt{2m}\left(\sqrt{E} - \sqrt{E - \hbar\omega}\right) \tag{4.22}$$

and

$$q_2 = \sqrt{2m}\left(\sqrt{E} + \sqrt{E - \hbar\omega}\right). \tag{4.23}$$

Let us observe that

$$\varepsilon(0, \hbar\omega) = \varepsilon_1 + i\varepsilon_2, \tag{4.24}$$

where

$$\varepsilon_1 = \operatorname{Re}[\varepsilon(0,\hbar\omega)] = n^2 - k^2, \quad (4.25)$$

$$\varepsilon_2 = \operatorname{Im}[\varepsilon(0,\hbar\omega)] = 2nk, \quad (4.26)$$

and where n is the coefficient of refraction and k the coefficient of extinction. Therefore,

$$\operatorname{Im}\left[\frac{-1}{\varepsilon(0,\hbar\omega)}\right] = \frac{\operatorname{Im}[\varepsilon(0,\hbar\omega)]}{|\varepsilon(0,\hbar\omega)|^2} = \frac{\varepsilon_2}{\varepsilon_1^2 + \varepsilon_2^2} = \frac{2nk}{(n^2+k^2)^2}. \quad (4.27)$$

Ashley [2, 3] showed that the relation between the dielectric function and the optical data may be approximated by

$$\operatorname{Im}\left[\frac{-1}{\varepsilon(q,\hbar\omega)}\right]$$

$$= \int_0^\infty d(\hbar\omega')\,\hbar\omega'\,\operatorname{Im}\left[\frac{-1}{\varepsilon(0,\hbar\omega')}\right]\delta\left[\hbar\omega - \left(\hbar\omega' + \frac{q^2}{2m}\right)\right]\frac{1}{\hbar\omega}. \quad (4.28)$$

As a consequence, for $E \geq 4E_f$ (where E_f is the Fermi energy), the stopping power may be calculated from

$$-\frac{dE}{ds} = \frac{me^2}{\pi\hbar^2 E}\int_0^{E/2} \operatorname{Im}\left[\frac{-1}{\varepsilon(0,\hbar\omega)}\right] G_e\left(\frac{\hbar\omega}{E}\right)\hbar\omega\,d(\hbar\omega). \quad (4.29)$$

Ashley also gave an accurate approximation for the function $G_e(x)$:

$$G_e(x)$$
$$= \ln\left(\frac{1.166}{x}\right) - \frac{3}{4}x - \frac{x}{4}\ln\left(\frac{4}{x}\right) + \frac{1}{2}x^{3/2} - \frac{x^2}{16}\ln\left(\frac{4}{x}\right) - \frac{31}{48}x^2. \quad (4.30)$$

4.3 Inelastic Mean Free Path

The electron inelastic mean free path can be calculated from the following equation:

$$\lambda_{inel} = \frac{1}{\int p(E,\hbar\omega)\,d(\hbar\omega)}, \quad (4.31)$$

where the integration is extended over all the allowed values of the energy transfer $\hbar\omega$. In a way similar to that used to calculate the stopping power, it is possible to show that

$$\lambda_{inel}^{-1} = \frac{me^2}{2\pi\hbar^2 E}\int_0^{E/2}\operatorname{Im}\left[\frac{-1}{\varepsilon(0,\hbar\omega)}\right]L_e\left(\frac{\hbar\omega}{E}\right)d(\hbar\omega), \quad (4.32)$$

where the approximate evaluation of $L_e(x)$ proposed by Ashley is

$$L_e(x) = (1-x)\ln\left(\frac{4}{x}\right) - \frac{7}{4}x + x^{3/2} - \frac{33}{32}x^2. \quad (4.33)$$

Note that, for the calculation of the inelastic cross-section of low-energy electrons, Powell [4] proposes the following differential cross-section:

$$\frac{d\lambda_{inel}^{-1}}{d(\hbar\omega)} = \frac{me^2}{2\pi\hbar^2 E}\mathrm{Im}\left[\frac{-1}{\varepsilon(0,\hbar\omega)}\right]\ln\left(\frac{cE}{\hbar\omega}\right), \quad (4.34)$$

where c is a parameter depending on the energy loss $\hbar\omega = \Delta E$.

4.4 Positrons

The stopping power and the inelastic mean free path for positrons may be calculated in a similar way [2, 3] (except for the exchange). For $4E_f < E < 10$ keV,

$$\left(-\frac{dE}{ds}\right)_p = \frac{me^2}{2\pi\hbar^2 E}\int_0^{E/2}\mathrm{Im}\left[\frac{-1}{\varepsilon(0,\hbar\omega)}\right]G_p\left(\frac{\hbar\omega}{E}\right)\hbar\omega\, d(\hbar\omega), \quad (4.35)$$

$$(\lambda_{inel}^{-1})_p = \frac{me^2}{2\pi\hbar^2 E}\int_0^{E/2}\mathrm{Im}\left[\frac{-1}{\varepsilon(0,\hbar\omega)}\right]L_p\left(\frac{\hbar\omega}{E}\right)d(\hbar\omega), \quad (4.36)$$

where

$$G_p(x) = \ln\left(\frac{1-x+\sqrt{1-2x}}{1-x-\sqrt{1-2x}}\right) \quad (4.37)$$

and

$$L_p(x) = \ln\left(\frac{1-x/2+\sqrt{1-2x}}{1-x/2-\sqrt{1-2x}}\right). \quad (4.38)$$

4.5 Plasma Oscillations

As plasmon excitation is very important in electron energy loss, we would like to have a closer look at this phenomenon. Let us give a brief treatment of the problem, considering the gas of the valence electrons in a metal: if n is the number of electrons per cubic centimetre, then the collective excitations of this gas produce oscillations as a result of the passage of electrons.

If ξ is a fluctuation of the distance r of the free-electron gas from the lattice of the positive charges of the solid, then the number ν of electrons in a spherical shell of "thickness" ξ is given by the following,

$$\nu = 4\pi r^2 n\xi, \quad (4.39)$$

so that the retarding force is

$$F = -\frac{e^2}{r^2}\nu = -4\pi e^2 n\xi. \quad (4.40)$$

As a consequence,

$$m\frac{d^2\xi}{dt^2} + 4\pi e^2 n\xi = 0,\tag{4.41}$$

and hence the motion is a harmonic oscillation of frequency ω_p, given by

$$\omega_p = \sqrt{\frac{4\pi e^2 n}{m}}.\tag{4.42}$$

A quantum of plasma oscillation is a plasmon. The energy of a plasmon is $E_p = \hbar\omega_p$. Plasma oscillations confined to a metal surface have a frequency ω_s, given by

$$\omega_s = \frac{\omega_p}{\sqrt{2}}.\tag{4.43}$$

The mean free path for plasmon emission can be calculated by using the Quinn formula [5],

$$\lambda_p = \frac{2a_0 E}{\omega_p}\left\{\ln\left[\frac{(p_f^2 + 2m\omega_p)^{1/2} - p_f}{p - (p^2 - 2m\omega_p)^{1/2}}\right]\right\}^{-1},\tag{4.44}$$

where

$$p = (2mE)^{1/2},\tag{4.45}$$

$$p_f = (2mE_f)^{1/2}\tag{4.46}$$

and E_f is the Fermi energy.

4.6 Comparing Theory and Experimental Data

The theoretical results that we present in Figs. 4.1–4.5 were obtained by following the method proposed by Ashley [2, 3] and using the mass absorption coefficients reported by Henke et al. [6].

The calculations of the refraction index n and of the extinction coefficient k were performed by use of [6]

$$n = 1 - \frac{e^2}{2\pi mc^2}\lambda^2 N \sum_p x_p f_{1p},\tag{4.47}$$

$$k = \frac{e^2}{2\pi mc^2}\lambda^2 N \sum_p x_p f_{2p}.\tag{4.48}$$

In these equations, c is the speed of light, N the number of molecules per unit volume (each having x_p atoms) and λ the photon wavelength. f_{1p} and f_{2p} are, respectively, the real and the imaginary component of the atomic scattering factor. The calculation of the real and imaginary component of the atomic scattering factor was performed using the following equations [6]:

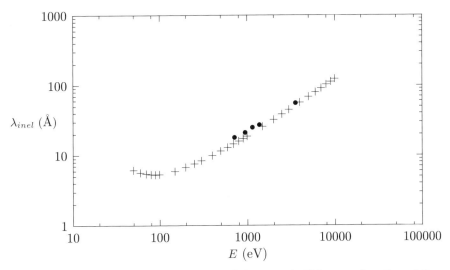

Fig. 4.1. Inelastic mean free path λ_{inel} of electrons in SiO_2 as a function of the electron kinetic energy E. +: theoretical calculation based on the optical data [7]. •: Flitsch and Raider experimental data [8]

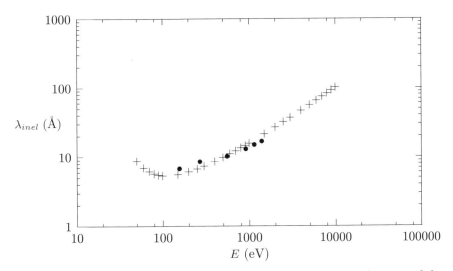

Fig. 4.2. Inelastic mean free path λ_{inel} of electrons in Al_2O_3 as a function of the electron kinetic energy E. +: theoretical calculation based on the optical data [7]. •: Battye et al. experimental data [9]

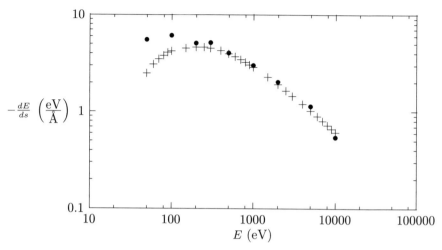

Fig. 4.3. Electron stopping power $-dE/ds$ in MgO as a function of the electron kinetic energy E. +: theoretical calculation based on the optical data [7]. •: experimental data reported by Akkerman et al. [10], taken from Joy's database [11]

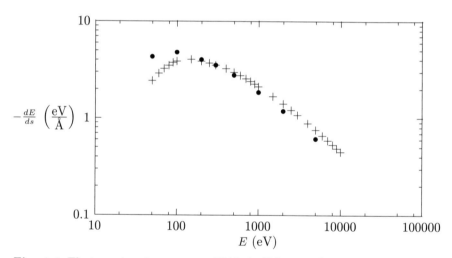

Fig. 4.4. Electron stopping power $-dE/ds$ in SiO_2 as a function of the electron kinetic energy E. +: theoretical calculation based on the optical data [7]. •: experimental data reported by Akkerman et al. [10], taken from Joy's database [11]

4.6 Comparing Theory and Experimental Data   51

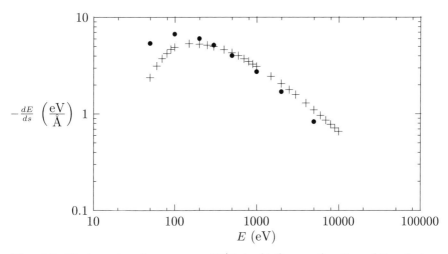

Fig. 4.5. Electron stopping power $-dE/ds$ in Al_2O_3 as a function of the electron kinetic energy E. $+$: theoretical calculation based on the optical data [7]. $\bullet$: experimental data reported by Akkerman et al. [10], taken from the Joy's database [11]

$$f_1 = Z + \frac{mc^2}{2\pi^2\hbar ce^2}\frac{A}{N_A}\int_0^\infty \frac{\epsilon^2\mu(\epsilon)}{E_p^2 - \epsilon^2}d\epsilon, \tag{4.49}$$

$$f_2 = \frac{mc^2}{4\pi\hbar ce^2}\frac{A}{N_A}E_p\mu(E_p). \tag{4.50}$$

In these equations E_p is the incident photon energy, Z the atomic number, A the atomic weight, N_A the Avogadro number and μ the photoabsorption cross-section (cm^2/g).

The Henke et al. data are given for discrete photon energies (beginning at 30.5 eV, and then 49.3 eV, 72.4 eV and so on), and we thus performed the calculation of the optical loss function utilising a cubic spline interpolation of the mass absorption coefficients. We performed polynomial extrapolation of the data to calculate the mass absorption coefficients for energies lower than 30.5 eV [7].

At high energy, the stopping power is well described by the Bethe formula. With $I = 142$ eV for SiO_2, the stopping power at 10 keV obtained with the Bethe formula is 0.458 eV/Å. The stopping power at 10 keV computed in the way described (i.e. using the optical data) is 0.462 eV/Å, 0.9% higher than Bethe's value.

The inaccuracy of the calculated inelastic mean free path and stopping power is less than $\sim 5 - 15\%$.

References

1. H.A. Bethe, in *Handbuch der Physik* (Springer, Berlin, 1933), **24**, 519
2. J.C. Ashley, J. Electron Spectrosc. Relat. Phenom. **46**, 199 (1988)
3. J.C. Ashley, J. Electron Spectrosc. Relat. Phenom. **50**, 323 (1990)
4. C.J. Powell, J. Vac. Sci. Technol. **13**, 219 (1976)
5. J. Quinn, Phys. Rev. **126**, 1453 (1962)
6. B.L. Henke, P. Lee, T.J. Tanaka, R.L. Shimabukuro, B.K. Fujikawa, At. Data Nucl. Data Tables **27**, 1 (1982)
7. M. Dapor, A. Miotello, Eur. Phys. J.: Appl. Phys. **5**, 143 (1999)
8. R. Flitsch, S.I. Raider, J. Vac. Sci. Technol. **12**, 305 (1975)
9. F.L. Battye, J.K. Jenkin, J. Liesegang, R.C.G. Leckey, Phys. Rev. B **9**, 2887 (1974)
10. A. Akkerman, T. Boutboul, A. Breskin, R. Chechik, A. Gibrekhterman, Y. Lifshitz, Phys. Status Solidi B **198**, 769 (1996)
11. D.C. Joy, Scanning **17**, 270 (1995)

5 Electrons Impinging on Solid Targets

5.1 Backscattered Electrons

The backscattering coefficient is the fraction of electrons of the primary beam that emerge from the surface of an electron-irradiated "bulk" target, where the meaning of the word "bulk" is explained in the following equations. The mean path length travelled by an electron in a solid target when its energy reduces from the primary energy E_0 to E is given by

$$s(E_0, E) = \int_0^s ds = \int_{E_0}^{E} \frac{dE}{dE/ds} \,. \tag{5.1}$$

This equation allows us to calculate the *maximum range* of penetration,

$$R(E_0) = \int_{E_0}^{E_{\min}} \frac{dE}{dE/ds} \,, \tag{5.2}$$

where $E_{\min}$ is the energy at which an electron can be considered absorbed by the target for any practical purpose.

When the thickness of the target is greater than $R(E_0)$, it is a bulk target for the primary electron energy considered. In general, when the thickness is lower than $R(E_0)$, the target is a film and the primary beam impinging on it is divided into three fractions, i.e. the backscattered, the absorbed and the transmitted electrons.

If the target is a bulk target, then the number of electrons transmitted is zero, while the fraction of electrons backscattered reaches its maximum value (the backscattering coefficient). The remaining primary particles are absorbed, so that, if r is the backscattering coefficient, the fraction of absorbed electrons equals $1 - r$.

Various semi-empirical approaches were proposed in the scientific literature in the early 1960s to treat the problem of electron backscattering. In 1960, Everhart [1] showed that the backscattering coefficient, for normal incidence, for high electron energy (greater than ∼10 keV) and for low–medium target atomic number (lower than ∼40–45), could be calculated from the following:

$$r = \frac{a - 1 + 0.5^a}{a + 1}, \tag{5.3}$$

where $a = 0.045Z$ and Z is the target atomic number.

The Everhart deduction is based on the semi-empirical Thomson–Whiddington energy loss relation [2]

$$E_0^2 - E^2 = cz, \tag{5.4}$$

where E represents the most probable electron beam energy at depth z in the target and c is a constant depending on the target. The Thomson–Whiddington energy loss relation can be obtained by assuming that the mean ionisation energy J depends linearly on E. The Bethe formula in such a case becomes

$$-\frac{dE}{dz} = \frac{k}{E}, \tag{5.5}$$

where k is a constant and, as a consequence,

$$z = -\int_{E_0}^{E} \frac{E\, dE}{k} = \frac{E_0^2}{2k} - \frac{E^2}{2k}. \tag{5.6}$$

With $c \equiv 2k$, this is the Thomson–Whiddington formula. Assuming that $E_{min} = 0$, then

$$E_0^2 = cR \tag{5.7}$$

and, as a consequence,

$$E^2 = c(R - z). \tag{5.8}$$

Let us introduce the supplement of the deflection angle, $\vartheta = \pi - \theta$. Taking into account the Thomson–Whiddington equation, the Rutherford cross-section for a pure Coulomb field, $d\sigma/d\Omega$, given by

$$\frac{d\sigma}{d\Omega} = \frac{Z^2 e^4}{4E^2} \frac{1}{(1 - \cos\theta)^2}, \tag{5.9}$$

becomes

$$d\sigma = \frac{Z^2 e^4}{4c} \frac{1}{R - z} \frac{2\pi \sin\vartheta\, d\vartheta}{(1 + \cos\vartheta)^2}. \tag{5.10}$$

The fractional range is defined as

$$u = \frac{z}{R}, \tag{5.11}$$

and the incremental number of electrons at a fractional depth u deflected through ϑ is given by

$$d\nu(\vartheta, u) = dN \frac{\nu}{S} d\sigma. \tag{5.12}$$

Here, N is the number of atoms per unit of volume in the target, so that $dN = NS\, dz$ gives the number of atoms in the volume $S\, dz$, while ν/S gives

5.1 Backscattered Electrons

the number of electrons incident on the surface, of area S. Once we have defined the quantity

$$a = \frac{\pi Z^2 e^4 N}{4c}, \tag{5.13}$$

the incremental number of electrons at a fractional depth u deflected through ϑ becomes

$$d\nu(\vartheta, u) = 2a \frac{\nu(u)du}{1-u} \frac{\sin\vartheta\, d\vartheta}{(1+\cos\vartheta)^2}. \tag{5.14}$$

On the plane at the fractional depth u, the number $\nu(u)$ of incident electrons can easily be calculated. Indeed,

$$\nu(u) = \nu_0 - \int_0^u \int_0^{\pi/2} d\nu(\vartheta, u'), \tag{5.15}$$

and therefore

$$\nu(u) = \nu_0 - a \int_0^u \frac{\nu(u')du'}{1-u'}. \tag{5.16}$$

The solution of this integral equation is as follows:

$$\nu(u) = \nu_0 (1-u)^a, \tag{5.17}$$

and hence

$$\frac{d\nu(\vartheta, u)}{\nu_0} = 2a(1-u)^{a-1} du \frac{\sin\vartheta\, d\vartheta}{(1+\cos\vartheta)^2}. \tag{5.18}$$

The total distance travelled by an electron reflected at an angle ϑ at depth z into the target is $s = z + z\sec\vartheta$. If ϑ_0 is the maximum escape angle for a given z, then $z(1 + \sec\vartheta_0) = R$, or $u(1 + \sec\vartheta_0) = 1$. With ϑ_0 satisfying the relationship

$$u(1 + \sec\vartheta_0) = 1, \tag{5.19}$$

the backscattering coefficient r is given by the following:

$$r = \int \frac{d\nu(\vartheta, u)}{\nu_0} = \int_0^{1/2} 2a(1-u)^{a-1} du \int_0^{\vartheta_0} \frac{\sin\vartheta\, d\vartheta}{(1+\cos\vartheta)^2}. \tag{5.20}$$

The integration over ϑ gives

$$\int_0^{\vartheta_0} \frac{\sin\vartheta\, d\vartheta}{(1+\cos\vartheta)^2} = \frac{1}{1+\cos\vartheta_0} - \frac{1}{2} = \frac{1}{2} - u \tag{5.21}$$

and, hence,

$$r = 2a \int_0^{1/2} (1-u)^{a-1} \left(\frac{1}{2} - u\right) du = \frac{a - 1 + (0.5)^a}{a+1}. \tag{5.22}$$

The simple Everhart theory can be used to obtain further useful information. In 1978, Jablonski [3] showed that it can be utilised to calculate the energy and angular distribution of backscattered electrons. As an electron reflected at depth z at an angle ϑ travels along a path given by

$$s = Ru(1 + \sec\vartheta) \tag{5.23}$$

and, in the Thomson–Whiddington approximation,

$$s = R\left[1 - \left(\frac{E}{E_0}\right)^2\right], \tag{5.24}$$

then

$$u = \frac{\cos\vartheta}{1 + \cos\vartheta}\left[1 - \left(\frac{E}{E_0}\right)^2\right]. \tag{5.25}$$

As a consequence,

$$du = \frac{2\cos\vartheta}{1 + \cos\vartheta}\frac{E\,dE}{E_0^2}. \tag{5.26}$$

If we define the function

$$w = 1 - \left(\frac{E}{E_0}\right)^2, \tag{5.27}$$

the energy and angular distribution of the backscattered electrons is then given by the following:

$$\frac{d^2\nu(\vartheta, E)}{dE\,d\vartheta} = \frac{4aE\nu_0}{E_0^2}\left(1 - \frac{\cos\vartheta}{1 + \cos\vartheta}w(E, E_0)\right)^{a-1}\frac{\sin\vartheta\cos\vartheta}{(1 + \cos\vartheta)^3}. \tag{5.28}$$

5.2 Electrons in thin films

5.2.1 Definitions, Symbols, Properties

Let us consider an electron beam impinging on a solid target. If R represents the maximum penetration range, the target is considered a film if its thickness s is smaller than R. If s is greater than R, the target is a bulk target. The fraction of particles transmitted in such a case is zero, while the fraction of particles backscattered is equal to the backscattering coefficient r (which depends on the target atomic number and on the electron primary energy).

Let us now concentrate our attention on an electron beam impinging on a film, i.e. $s \leq R$, so that the beam is split into three fractions η_A, η_B and η_T representing the fraction of particles absorbed, backscattered and transmitted, respectively. Each of them lies in the range $[0, 1]$. Since the total number of particles has to be conserved, then $\eta_A + \eta_B + \eta_T = 1$ for any given thickness.

Note that if the film is deposited on a substrate, the fractions mentioned above are different from those corresponding to unsupported thin films. The

5.2 Electrons in thin films

difference is due to the backscattering from the substrate. In order to distinguish supported and unsupported thin films, we shall use the symbols ζ_A, ζ_B and ζ_T for the absorbed, backscattered and transmitted fractions, respectively, for supported thin films. Note that each fraction ζ lies in the range $[0, 1]$. Conservation of the number of particles for the fractions ζ reads $\zeta_A + \zeta_B + \zeta_T = 1$. Obviously, owing to the backscattering from the substrate, $\zeta_B > \eta_B$ and $\zeta_T < \eta_T$ for any given film thickness.

The fractions η and ζ depend on $\xi(s)$, a function related to the scattering processes. A rough evaluation of $\xi(s)$ is the following:

$$\xi(s) = N\sigma_{tr}s, \tag{5.29}$$

where N is the number of atoms per unit of volume in the target and σ_{tr} is the elastic transport cross-section.

Let us observe that

$$\lim_{\xi \to 0} \eta_A(\xi) = 0, \tag{5.30}$$

$$\lim_{\xi \to 0} \eta_B(\xi) = 0, \tag{5.31}$$

$$\lim_{\xi \to 0} \eta_T(\xi) = 1, \tag{5.32}$$

$$\lim_{\xi \to \infty} \eta_A(\xi) = 1 - r, \tag{5.33}$$

$$\lim_{\xi \to \infty} \eta_B(\xi) = r, \tag{5.34}$$

$$\lim_{\xi \to \infty} \eta_T(\xi) = 0. \tag{5.35}$$

Let us now consider an electron beam impinging on a film of a given material x of thickness s which has been deposited on a substrate composed of a material y (different from material x). R_x indicates the maximum penetration range in a bulk target composed of x, and R_y the maximum penetration range in a bulk target composed of y. Let us assume that the thickness of the substrate is greater than R_y, so that the substrate is a bulk target for the primary energy considered.

The maximum penetration range is a combination of R_x and R_y whose value depends on the film thickness. It approaches R_y for $s \to 0$ and R_x for $s \to R_x$.

Indicating by r_x the backscattering coefficient of x and by r_y the backscattering coefficient of y, the following conditions have to be satisfied:

$$\lim_{\xi \to 0} \zeta_A(\xi) = 0, \tag{5.36}$$

$$\lim_{\xi \to 0} \zeta_B(\xi) = r_y, \tag{5.37}$$

$$\lim_{\xi \to 0} \zeta_T(\xi) = 1 - r_y, \tag{5.38}$$

$$\lim_{\xi \to \infty} \zeta_A(\xi) = 1 - r_x , \tag{5.39}$$

$$\lim_{\xi \to \infty} \zeta_B(\xi) = r_x , \tag{5.40}$$

$$\lim_{\xi \to \infty} \zeta_T(\xi) = 0 . \tag{5.41}$$

5.2.2 Unsupported thin films

Let us indicate by $\Delta \xi$ an increment of ξ, i.e.

$$\Delta \xi = \xi(s + \Delta s) - \xi(s) , \tag{5.42}$$

and let us calculate the fraction of particles absorbed by an unsupported thin film of thickness $s + \Delta s$. This is given by the fraction absorbed by ξ plus the fraction transmitted through ξ and absorbed by $\Delta \xi$, plus the fraction transmitted through ξ, backscattered by $\Delta \xi$ and absorbed by ξ, and so on, in a sequence of infinite reflections:

$$\begin{aligned}
\eta_A[\xi(s+\Delta s)] &= \eta_A(\xi + \Delta \xi) \\
&= \eta_A(\xi) + \eta_T(\xi) \eta_B(\Delta \xi) \eta_A(\xi) \sum_{n=0}^{\infty} [\eta_B(\xi) \eta_B(\Delta \xi)]^n \\
&\quad + \eta_T(\xi) \eta_A(\Delta \xi) \sum_{n=0}^{\infty} [\eta_B(\xi) \eta_B(\Delta \xi)]^n \\
&= \eta_A(\xi) + \frac{\eta_A(\xi) \eta_T(\xi) \eta_B(\Delta \xi) + \eta_T(\xi) \eta_A(\Delta \xi)}{1 - \eta_B(\xi) \eta_B(\Delta \xi)} .
\end{aligned} \tag{5.43}$$

Proceeding in a similar way, we obtain the following:

$$\eta_B(\xi + \Delta \xi) = \eta_B(\xi) + \frac{\eta_T^2(\xi) \eta_B(\Delta \xi)}{1 - \eta_B(\xi) \eta_B(\Delta \xi)} , \tag{5.44}$$

$$\eta_T(\xi + \Delta \xi) = \frac{\eta_T(\xi) \eta_T(\Delta \xi)}{1 - \eta_B(\xi) \eta_B(\Delta \xi)} . \tag{5.45}$$

For $\xi \to \infty$, $\eta_A(\xi) \to 1 - r$, $\eta_B(\xi) \to r$ and $\eta_T(\xi) \to 0$. Then, as a consequence, for $\Delta \xi \to \infty$, $\eta_A(\xi + \Delta \xi) \to 1 - r$, $\eta_A(\Delta \xi) \to 1 - r$ and $\eta_B(\Delta \xi) \to r$.

Let us define

$$\mu = \frac{1 + r^2}{2r} , \tag{5.46}$$

and

$$\nu = \frac{1 - r^2}{2r} , \tag{5.47}$$

Using (5.43) for $\Delta\xi \to \infty$, it is possible to see that

$$\eta_T^2 = \eta_B^2 - 2\mu\eta_B + 1 \ . \tag{5.48}$$

Let us introduce β, the derivative of η_B calculated for $\xi = 0$:

$$\beta \equiv \lim_{\xi \to 0} \left[\frac{\eta_B(\xi)}{\xi}\right] = \frac{d\eta_B(\xi)}{d\xi}\Big|_{\xi=0} \ . \tag{5.49}$$

From (5.44), we obtain the following:

$$\frac{\eta_B(\xi + \Delta\xi) - \eta_B(\xi)}{\Delta\xi} = \frac{\eta_B(\Delta\xi)}{\Delta\xi} \frac{\eta_T^2(\xi)}{1 - \eta_B(\xi)\eta_B(\Delta\xi)} \ . \tag{5.50}$$

Therefore

$$\frac{d\eta_B(\xi)}{d\xi} = \beta\eta_T^2(\xi) = \beta[\eta_B^2(\xi) - 2\mu\eta_B(\xi) + 1] \ . \tag{5.51}$$

This equation is equivalent to

$$\frac{d\eta_B}{\eta_B - (1/r)} - \frac{d\eta_B}{\eta_B - r} = 2\nu\beta\,d\xi \ . \tag{5.52}$$

Remembering that $\eta_B(0) = 0$, the integration of this equation gives

$$\eta_B(\xi) = r\frac{1 - \exp(-2\nu\beta\xi)}{1 - r^2\exp(-2\nu\beta\xi)} \ . \tag{5.53}$$

As $\eta_T^2 = \eta_B^2 - 2\mu\eta_B + 1$ and $\eta_A = 1 - \eta_B - \eta_T$, we can conclude that

$$\eta_T(\xi) = \frac{1 - r^2}{1 - r^2\exp(-2\nu\beta\xi)}\exp(-\nu\beta\xi) \tag{5.54}$$

and

$$\eta_A(\xi) = (1 - r)\frac{r\exp(-2\nu\beta\xi) - (1 + r)\exp(-\nu\beta\xi) + 1}{1 - r^2\exp(-2\nu\beta\xi)} \ . \tag{5.55}$$

Equations (5.53), (5.54) and (5.55) have been taken from H.W. Schmidt [4].

5.2.3 Supported thin films

Let us now consider a supported thin film x deposited on a substrate y and remember that ζ_A is the fraction of electrons absorbed in the supported thin film, ζ_B the fraction of electrons backscattered and ζ_T the fraction of electrons transmitted across the interface between the film x and substrate y. To calculate ζ_A, let us observe that the fraction η_A is absorbed by the film, and then the fraction $\eta_T r_y$ comes back across the interface so that the fraction $\eta_T r_y \eta_A$ is absorbed. From the infinite sum of contributions to the absorbed fraction ζ_A, we obtain

$$\zeta_A = \eta_A + \eta_T r_y \eta_A + \eta_T r_y \eta_B r_y \eta_A + \cdots$$

$$= \eta_A \left[1 + \eta_T r_y \sum_{n=0}^{\infty} (\eta_B r_y)^n \right]$$

$$= \eta_A \left(1 + \frac{\eta_T r_y}{1 - \eta_B r_y} \right) . \tag{5.56}$$

In a similar way, we obtain the following, for ζ_B:

$$\zeta_B = \eta_B + \eta_T^2 r_y \sum_{n=0}^{\infty} (\eta_B r_y)^n = \eta_B + \frac{\eta_T^2 r_y}{1 - \eta_B r_y} . \tag{5.57}$$

We obtain the following for ζ_T:

$$\zeta_T = \eta_T \sum_{n=0}^{\infty} (\eta_B r_y)^n - \eta_T r_y \sum_{n=0}^{\infty} (\eta_B r_y)^n = \frac{\eta_T (1 - r_y)}{1 - \eta_B r_y} . \tag{5.58}$$

If the material of the film is the same as that of the substrate, we can avoid the subscripts x and y in the equations above, so we can write

$$\zeta_A = \eta_A \left(1 + \frac{\eta_T r}{1 - \eta_B r} \right) , \tag{5.59}$$

$$\zeta_B = \eta_B + \frac{\eta_T^2 r}{1 - \eta_B r} = r , \tag{5.60}$$

$$\zeta_T = \eta_T \frac{1 - r}{1 - \eta_B r} . \tag{5.61}$$

Upon substituting the Schmidt equations (5.53), (5.54) and (5.55) into the previous equations, we obtain the following [5]:

$$\zeta_A = (1 - r)[1 - \exp(-\nu \beta \xi)] , \tag{5.62}$$

$$\zeta_B = r , \tag{5.63}$$

$$\zeta_T = (1 - r) \exp(-\nu \beta \xi) . \tag{5.64}$$

Let us observe that the derivative of ζ_A gives the implantation profile, or the depth distribution, of the trapped electrons [5]:

$$\frac{d\zeta_A}{ds} = \nu \beta (1 - r) \exp(-\nu \beta \xi) \frac{d\xi}{ds} . \tag{5.65}$$

Note that the relationship (5.48) between η_T and η_B for unsupported thin films of the material x, taking into account the definition of μ (5.46), can also be rewritten as

$$\eta_T = \sqrt{\frac{(r_x - \eta_B)(1 - r_x \eta_B)}{r_x}} . \tag{5.66}$$

Substituting (5.66) in (5.57), we can conclude that the fraction of backscattered particles is given by the following, for a thin film of the material x supported on a substrate of the material y [6]:

$$\zeta_B = \eta_B + (r_x - \eta_B)\frac{r_y}{r_x}\frac{1 - r_x \eta_B}{1 - r_y \eta_B} \ . \tag{5.67}$$

All the equations deduced in this section are based on a multiple reflection method and were derived neglecting the fact that the absorption, backscattering and transmission coefficients change for a given layer owing to the change in the energy and angular distribution of the particles that repeatedly are incident on the layer. Hence these equations represent an approximate approach. However, the equations are certainly useful for gaining insight into the subject of the influence of backscattering from the substrate and also to understand the theoretical framework.

As a consequence, the present theory may be informative, but one must keep in mind the limits of the analytical procedure. The fact that (5.67) is a simple, closed formula certainly renders it attractive. Once the backscattering coefficients of the two materials and the fraction of particles backscattered by an unsupported thin film are given, the theory allows one to calculate the fraction of backscattered particles when the same film is deposited on a substrate.

5.3 Secondary Electrons

Electron beams impinging on solid targets stimulate the emission of secondary electrons, i.e. electrons extracted from the atoms bound in the solid, as a consequence of interaction with electrons of the primary beam or with the other energetic secondary electrons that travel in the solid.

Some secondary electrons, following a number of elastic and inelastic collisions with the atoms of the solid, reach the surface of the solid with enough energy to emerge from it. Even some of the primary electrons, after several elastic and inelastic collisions with the target atoms, emerge from the surface. So the spectrum of the secondary electrons is *contaminated* by the contribution of the backscattered primary electrons. In the following, we shall concentrate our attention on the secondary electrons, choosing to neglect the backscattered electrons already treated in the previous sections. In particular, we shall describe the theory of the secondary-electron cascade in metals as proposed by Wolff in 1954 [7].

The process of secondary-electron emission can be conceptually divided into two phenomena: the first concerns the production of secondary electrons due to interaction between primary electrons and electrons bound in the solid; the second is represented by the *cascade*, whereby the secondary electrons, diffusing in the solid, extract new secondary electrons in a cascade process.

5 Electrons Impinging on Solid Targets

Since every secondary electron loses energy during its travel within the solid, the process proceeds until the secondary-electron energy has decreased so that it is not sufficient to extract new secondary electrons or until the electron reaches the surface of the solid with sufficient energy to emerge from it.

Let us introduce some relevant physical quantities. We shall indicate by $\mathcal{N}(\boldsymbol{r}, \boldsymbol{\Omega}, E, t)$ the number of electrons at time t between $\boldsymbol{r}$ and $\boldsymbol{r} + d\boldsymbol{r}$, between $\boldsymbol{\Omega}$ and $\boldsymbol{\Omega} + d\boldsymbol{\Omega}$, and between E and $E + dE$, where $\boldsymbol{\Omega}$ represents a unit vector in the direction of the velocity of the electron $\boldsymbol{v}$. $\lambda(E)$ is the electron mean free path and $F(\boldsymbol{\Omega}, E; \boldsymbol{\Omega}', E')$ the probability that an electron at $\boldsymbol{\Omega}', E'$ is found, after scattering, at $\boldsymbol{\Omega}, E$. Let us indicate by $S(\boldsymbol{r}, \boldsymbol{\Omega}, E, t)$ the source, i.e. the density of the secondary electrons due to the bombardment of primary particles. The electron cascade is governed by the Boltzmann diffusion equation,

$$\frac{\partial \mathcal{N}}{\partial t} + \boldsymbol{v} \cdot \boldsymbol{\nabla} \mathcal{N} = -\frac{v\mathcal{N}}{\lambda} + S$$
$$+ \int dE' d\Omega' \frac{v'\mathcal{N}(\boldsymbol{r}, \boldsymbol{\Omega}', E', t)}{\lambda(E')} F(\boldsymbol{\Omega}, E; \boldsymbol{\Omega}', E') \,. \qquad (5.68)$$

Considering a geometry in which the primary particles are incident normally on the target surface, the problem has azimuthal symmetry and involves the distance to the surface z, the energy E, and the angle between the velocity of the secondary electron and the direction normal to the surface, θ. If we concentrate our attention on the steady-state condition

$$\frac{\partial \mathcal{N}}{\partial t} = 0 \,, \qquad (5.69)$$

then

$$\boldsymbol{v} \cdot \boldsymbol{\nabla} \mathcal{N} = -\frac{v\mathcal{N}}{\lambda} + S$$
$$+ \int dE' d\Omega' \frac{v'\mathcal{N}(\boldsymbol{r}, \boldsymbol{\Omega}', E', t)}{\lambda(E')} F(\boldsymbol{\Omega}, E; \boldsymbol{\Omega}', E') \,. \qquad (5.70)$$

Let us indicate by Θ the angle between $\boldsymbol{\Omega}$ and $\boldsymbol{\Omega}'$ and expand in spherical harmonics the three functions $\mathcal{N}$, F and S:

$$\mathcal{N}(z, \cos\theta, E) = \frac{1}{4\pi} \sum_{l=0}^{\infty} (2l+1) \mathcal{N}_l(z, E) P_l(\cos\theta) \,, \qquad (5.71)$$

$$S(z, \cos\theta, E) = \frac{1}{4\pi} \sum_{l=0}^{\infty} (2l+1) S_l(z, E) P_l(\cos\theta) \,, \qquad (5.72)$$

$$F(\boldsymbol{\Omega}, E; \boldsymbol{\Omega}', E') = F(\cos\Theta; E, E')$$
$$= \frac{1}{4\pi} \sum_{l=0}^{\infty} (2l+1) F_l(E, E') P_l(\cos\Theta) \,. \qquad (5.73)$$

Let us introduce the function

$$\psi_l = v\mathcal{N}_l/\lambda(E),\tag{5.74}$$

and note that

$$\begin{aligned}
\mathbf{v}\cdot\nabla\mathcal{N} &= \frac{1}{4\pi}\sum_l(2l+1)v\frac{\partial\mathcal{N}_l}{\partial z}\cos\theta\,P_l(\cos\theta)\\
&= \frac{\lambda(E)}{4\pi}\sum_l(2l+1)\frac{\partial}{\partial z}\left(\frac{v\mathcal{N}_l}{\lambda}\right)\cos\theta\,P_l(\cos\theta)\\
&= \frac{\lambda}{4\pi}\sum_l(2l+1)\frac{\partial\psi_l}{\partial z}\cos\theta\,P_l(\cos\theta)\\
&= \frac{\lambda}{4\pi}\sum_l\frac{\partial\psi_l}{\partial z}[(l+1)P_{l+1}(\cos\theta)+lP_{l-1}(\cos\theta)]\\
&= \frac{\lambda}{4\pi}\sum_l\frac{\partial\psi_{l-1}}{\partial z}lP_l(\cos\theta)+\frac{\lambda}{4\pi}\sum_l\frac{\partial\psi_{l+1}}{\partial z}(l+1)P_l(\cos\theta)\\
&= \frac{1}{4\pi}\sum_l(2l+1)\lambda(E)\left(\frac{l}{2l+1}\frac{\partial\psi_{l-1}}{\partial z}+\frac{l+1}{2l+1}\frac{\partial\psi_{l+1}}{\partial z}\right)P_l(\cos\theta).
\end{aligned}$$
(5.75)

Then,

$$\psi_l(z,E) = S_l(z,E)+\int dE'\,\psi_l(z,E')F_l(E,E')$$

$$+\lambda(E)\left(\frac{l}{2l+1}\frac{\partial\psi_{l-1}}{\partial z}+\frac{l+1}{2l+1}\frac{\partial\psi_{l+1}}{\partial z}\right).\tag{5.76}$$

If we suppose that the secondary electrons are uniformly produced throughout the depth, the functions ψ_l depend only on E and are independent of z. If the energy is sufficiently low, the distribution of the secondary electrons is spherically symmetric, so one can neglect all the harmonics higher than $l=0$. Within this approximation, the integro-differential equation (5.76) with $S_0=0$ becomes

$$\psi_0(E) = \int_E^\infty dE'\,\psi_0(E')F_0(E,E').\tag{5.77}$$

Note that, once ψ_0 is known, the secondary-electron energy distribution can be calculated as

$$j(E) = \mathcal{N}_0 v = \psi_0\lambda(E).\tag{5.78}$$

Concerning $F_0(E, E')$, let us note that

$$2\pi \int \sin\Theta\, d\Theta\, F(\cos\Theta; E, E') P_k(\cos\Theta)$$

$$= 2\pi \int_0^\pi \sin\Theta\, d\Theta \frac{1}{4\pi} \sum_{l=0}^\infty (2l+1) F_l(E, E') P_l(\cos\Theta) P_k(\cos\Theta)$$

$$= \frac{1}{2} \sum_{l=0}^\infty (2l+1) F_l(E, E') \int_{-1}^1 P_l(u) P_k(u)\, du$$

$$= \frac{1}{2} \sum_{l=0}^\infty (2l+1) F_l(E, E') \frac{2}{2l+1} \delta_{lk}$$

$$= F_k(E, E'), \tag{5.79}$$

so that

$$F_0(E, E') = 2\pi \int \sin\Theta\, d\Theta\, F(\cos\Theta; E, E'). \tag{5.80}$$

Therefore, within the approximation of spherical symmetry of the secondary-electron distribution and ignoring all the harmonics higher than $l = 0$, (5.77) can also be written as the following:

$$\psi_0(E) = 2\pi \int_E^\infty dE'\, \psi_0(E') \int_0^\pi \sin\Theta\, d\Theta\, F(\cos\Theta; E, E'). \tag{5.81}$$

Here $F_0(E, E')$ represents the total probability of scattering between the energies E and E', independently of the angle.

Let us assume that, on average, electrons of energy $E' < 100$ eV lose half of their energy at each collision. Let us introduce the average energy after scattering, $\bar{E}$, defined by

$$\bar{E} = \gamma E'. \tag{5.82}$$

Here E' is the energy before the scattering and, for energies four times greater than the Fermi energy, γ is $\sim 1/2$. Within this approximation,

$$F_0(E, E') = 2\,\delta(E - \bar{E}) = 2\,\delta(E - \gamma E'). \tag{5.83}$$

Note the factor of 2, which takes into account the fact that there are two electrons after each electron collides.

Equation (5.77) therefore becomes

$$\psi_0(E) = 2 \int \delta(E - \gamma E') \psi_0(E')\, dE', \tag{5.84}$$

which, as the reader can easily verify, produces the following solution:

$$\psi_0(E) = \frac{2}{\gamma(E)} \psi_0\left[\frac{E}{\gamma(E)}\right]. \tag{5.85}$$

Let us now define the function $x(E)$, so that

$$2[\gamma(E)]^{x(E)-1} = 1 . \tag{5.86}$$

Our solution is proportional to E^{-x}:

$$\psi_0(E) \propto E^{-x(E)} . \tag{5.87}$$

Assuming that the validity of the Wolff theory can be extended up to ~1000 eV, the semi-empirical law proposed in 1977 by Sickafus [8, 9, 10] immediately follows. As the inelastic mean free path $\lambda(E)$ is approximately proportional to $E/\log E$, i.e.

$$\lambda \propto \frac{E}{\log E} , \tag{5.88}$$

we can see that the secondary-electron energy distribution is approximately given by

$$j(E) = \psi_0 \lambda(E) = \frac{C}{E^{x-1} \log E} = \frac{C E^{-m}}{\log E} , \tag{5.89}$$

where C is a constant and $m = x - 1$. Then,

$$\log j(E) = \log C - \log(\log E) - m \log E . \tag{5.90}$$

Note that the function $\log(\log E)$ is almost constant. As a result, in the range of energies 10–1000 eV above the low-energy cascade peak,

$$j(E) = A E^{-m} , \tag{5.91}$$

where A is almost constant and $m \sim 1$.

5.4 Comparing Theory and Experimental Data

We have introduced many physical quantities and described them through semi-empirical approaches in this chapter. Today, the most accurate theoretical calculation of these physical quantities can be performed by Monte Carlo simulations, which will be faced in the next chapter. Consequently, the present chapter shows only a comparison of the results obtained from the multiple reflection method with the experimental data available concerning backscattering from supported thin films. This has been done because, despite the simplicity of the approach, there is substantial agreement between theory and experiment. The other quantities introduced in this chapter (backscattering coefficient, depth distribution of the trapped electrons and positrons, secondary-electron emission, and so on) will be calculated in the next chapter by introducing the elastic and inelastic cross-sections (described in the previous chapters) into the Monte Carlo simulations.

In order to investigate the applicability of the multiple reflection method and to compare its results with the experimental data reported by Niedrig in

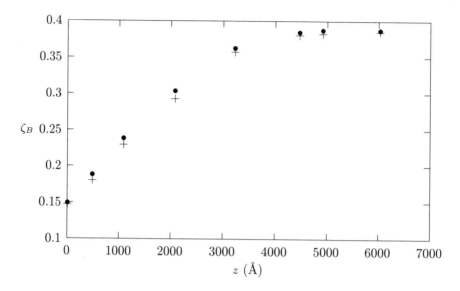

Fig. 5.1. Electron backscattering ratio ζ_B of Ag surface films deposited on Al bulk substrates versus surface film thickness z. The data presented concern 20 keV electron beams irradiating the targets in the $+z$ direction. $+$: theoretical calculation based on the multiple reflection method (5.67) [6]. •: Niedrig's experimental data [11]

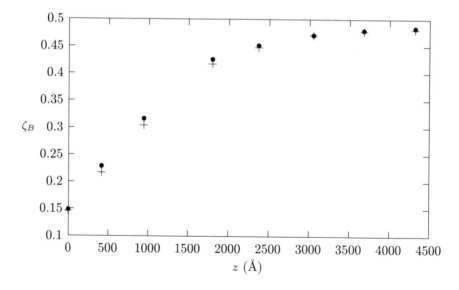

Fig. 5.2. Electron backscattering ratio ζ_B of Au surface films deposited on Al bulk substrates versus surface film thickness z. The data presented concern 20 keV electron beams irradiating the targets in the $+z$ direction. $+$: theoretical calculation based on the multiple reflection method (5.67) [6]. •: Niedrig's experimental data [11]

5.4 Comparing Theory and Experimental Data   67

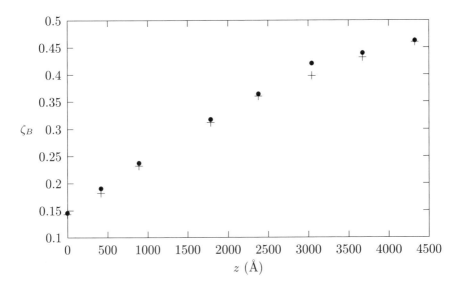

Fig. 5.3. Electron backscattering ratio ζ_B of Au surface films deposited on Al bulk substrates versus surface film thickness z. The data presented concern 30 keV electron beams irradiating the targets in the $+z$ direction. +: theoretical calculation based on the multiple reflection method (5.67) [6]. •: Niedrig's experimental data [11]

[11] we used Niedrig's experimental data for the values of $f_B(z)$, r_t and r_s in (5.67). The agreement between the theoretical and the experimental values of $\zeta_B(z)$ is excellent (see Figs. 5.1–5.3).

Note that (5.67) was derived by the multiple reflection method, assuming that fixed values for the probabilities of absorption, backscattering and transmission could be attributed to both the surface layer and the substrate. However, this is not strictly valid, as these probabilities depend on the angular distribution of the particles and on their energies. Since the particles dissipate energy in ionisation, electron excitation, plasmon emission and so on, the assumption of fixed probabilities represents an approximation [12]: our comparisons were motivated by the necessity to evaluate the accuracy of such an approximation.

Despite the approximations involved, the agreement, both in the trend and in the absolute values, with Niedrig's experimental data concerning the values of $\zeta_B(z)$ shows that the multiple reflection method is a reliable approach for calculations of backscattering from supported thin films and that it is useful for fast and quite accurate evaluations when data concerning unsupported thin films and backscattering coefficients are available.

In conclusion, the accord that we found indicates that (5.67) can be safely used for calculating backscattering from supported films and that the multiple reflection method is rather a good approximation.

References

1. T. Everhart, J. Appl. Phys. **31**, 1483 (1960)
2. J.J. Thomson, *Conduction of Electricity Through Gases* (Cambridge University Press, Cambridge, England, 1906), 2nd ed.
3. A. Jablonski, Surf. Sci. **74**, 621 (1978)
4. H.W. Schmidt, Ann. Phys. (Leipzig) **23**, 671 (1907)
5. M. Dapor, Phys. Rev. B **43**, 10118 (1991); Phys. Rev. B **44**, 9784 (1991)
6. M. Dapor, Eur. Phys. J.: Appl. Phys. **18**, 155 (2002)
7. P.A. Wolff, Phys. Rev. **95**, 56 (1954)
8. E.N. Sickafus, Phys. Rev. B **16**, 1436 (1977)
9. E.N. Sickafus, Phys. Rev. B **16**, 1448 (1977)
10. E.N. Sickafus, C. Kukla, Phys. Rev. B **19**, 4056 (1979)
11. H. Niedrig, J. Appl. Phys. **53**, R15 (1982)
12. D. Liljequist, J. Phys. D: Appl. Phys. **10**, 1363 (1977)

6 Monte Carlo Simulations

6.1 The Monte Carlo Method

The study of the physical quantities that are involved in electron–solid interaction (backscattering coefficient, energy and angular distribution of backscattered electrons, electron penetration in thin solid films, electron implantation profiles in bulk targets, secondary-electron energy distribution, etc.) requires approximations and approaches which are valid only for limited ranges of energy and angle and for selected materials. All the analytical approaches we have proposed involve approximations and have been deduced using semi-empirical models.

A completely different approach to the determination of the quantities mentioned above is represented by the Monte Carlo method. Within statistical uncertainty, which can be reduced arbitrarily, this method of simulating electron–solid interactions is very accurate and gives results in excellent agreement with the experimental data available.

The Monte Carlo method is a mathematical tool utilising random numbers. This method can obtain accurate results about the absorption, backscattering and transmission of electrons penetrating both supported and unsupported thin films, and can also provide implantation profiles of the absorbed electrons, the energy and angular distributions of backscattered and transmitted electrons, and spectra of secondary electrons. The cross-sections and probabilistic laws for interaction of electrons with the atoms constituting the target are known, so that they can be used in the Monte Carlo code.

When a stream of electrons impinges on a solid target, the incident particles (during their travel within the solid) lose energy and change direction at each collision with the atoms bound in the solid. Nuclear collisions deflect the electrons without relevant kinetic-energy transfer, owing to the large difference between the masses of the electron and the nucleus: the differential elastic scattering cross-section has to be used in order to describe these processes (see Chap. 3).

On the other hand, excitations and ejections of atomic electrons, and plasmon excitations affect the energy dissipation but only slightly affect the direction of the electron in the solid. These can be described as inelastic processes that are essentially governed by the equations for the stopping power and/or the inelastic cross-section (see Chap. 4).

All the cross-sections, mean free paths and stopping powers can thus be accurately calculated by the Monte Carlo method in order to obtain the macroscopic characteristics of the interaction processes (see Chap. 5), by simulating a large number of single particle trajectories and then averaging them.

However, as the Monte Carlo method is statistical, the accuracy of its results depends on the number of simulated trajectories. Recent evolution in computer calculation capability means we are now able to obtain statistically significant results in very short times of calculation.

6.2 Random Variables

We need to introduce the concept of a random variable in order to understand the Monte Carlo method. Let us begin with the discrete random variable. In order to specify a discrete random variable ξ, one has to provide the n values that ξ can assume, with their probabilities of occurrence.

In other words, we have to specify $\xi_1, \xi_2, \ldots, \xi_n$ and their probabilities $p_1 = P(\xi = \xi_1), p_2 = P(\xi = \xi_2), \ldots, p_n = P(\xi = \xi_n)$. Here, we have indicated by $P(\xi = \xi_i)$ the probability p_i that a value of ξ chosen at random is equal to ξ_i, with $i = 1, 2, \ldots, n$. Of course, $\sum_{i=1}^{n} p_i = 1$, and for every $i = 1, 2, \ldots, n$, $p_i \geq 0$. Once a discrete random variable has been specified, its expected value can be calculated from

$$\langle \xi \rangle = \frac{\sum_i^n \xi_i p_i}{\sum_i^n p_i} = \sum_i^n \xi_i p_i \ . \tag{6.1}$$

In order to define a continuous random variable ξ, one has to give the range of variability and the distribution of the probabilities. If a continuous random variable ξ is defined, this means that we have specified a real interval (a, b) and a probability density $p(x)$ of the distribution of the variable. Let us now consider the real interval (c, d) in (a, b). If $P(c < \xi < d)$ indicates the probability that a value of ξ taken at random falls in the interval (c, d), then

$$P(c < \xi < d) = \int_c^d p(x)\, dx \ . \tag{6.2}$$

As $\int_a^b p(x)\, dx = 1$, the expected value of ξ can be calculated from

$$\langle \xi \rangle = \int_a^b x p(x)\, dx \ . \tag{6.3}$$

If we need to calculate the expected value of a function $f(\xi)$, this is given by

$$\langle f(\xi) \rangle = \int_a^b f(x) p(x)\, dx \ . \tag{6.4}$$

6.2.1 Random Variable Uniformly Distributed in the Interval (0, 1)

A random variable μ is said to be uniformly distributed in the interval (0, 1) if it is defined in the range (0, 1) and if its probability density is simply given by $p_\mu(x)=1$.

The probability that a random variable uniformly distributed in the interval (0, 1) assumes a value between c and d (where $c > 0$ and $d < 1$) is the length of the interval. In fact,

$$P(c < \mu < d) = \int_c^d p_\mu(x)\, dx = \int_c^d dx = d - c\,. \tag{6.5}$$

The expected value of a random variable uniformly distributed in the interval (0, 1) equals $1/2$, as follows:

$$\langle \mu \rangle = \int_0^1 x p_\mu(x)\, dx = \int_0^1 x\, dx = \frac{1}{2}\,. \tag{6.6}$$

6.2.2 Random Variable Distributed in a Given Interval with a Given Probability

If ξ indicates a random variable defined in the interval (a, b) and distributed with a given probability density $p(x)$, and μ a random variable uniformly distributed in the range (0, 1), the values of ξ are related to those of μ and can be obtained from the equation

$$\int_a^\xi p(x)\, dx = \mu\,. \tag{6.7}$$

Let us consider the function $f(x)$ defined by

$$f(x) = \int_a^x p(z)\, dz\,. \tag{6.8}$$

The function $f(x)$ has the following properties: it increases when x increases because $f(a) = 0$ and $f(b) = 1$ and, by the definition of probability, $df(x)/dx = p(x) > 0$.

To proceed, let us consider the straight line

$$y = \mu\,, \tag{6.9}$$

where μ is a real number in the interval (0, 1). The line is parallel to the x axis and must intersect the increasing function $f(x)$ in only one point. Therefore, a intersection between the two curves exists and occurs only once, i.e. for each value of μ, (6.7) has a solution, and this solution is unique.

Let us now show that the distribution ξ defined as described here has a probability distribution $p(x)$. If we consider an interval (c, d) contained in the interval (a, b), then

$$P(c < \xi < d) = P[f(c) < \mu < f(d)]\,. \tag{6.10}$$

As the probability density of μ equals 1, then
$$P[f(c) < \mu < f(d)] = f(d) - f(c) \tag{6.11}$$
and, on the other hand,
$$f(d) - f(c) = \int_c^d p(x)\,dx. \tag{6.12}$$
From the last three equations, we obtain the result that
$$P(c < \xi < d) = \int_c^d p(x)\,dx, \tag{6.13}$$
and hence the random variable ξ is distributed with a probability density $p(x)$ in the interval (a, b).

6.2.3 Random Variable Uniformly Distributed in the Interval (a, b)

We are now interested in calculating a random variable η uniformly distributed in the real interval (a, b), which is useful in Monte Carlo simulations of electron–solid interactions. The requirement that the distribution is uniform implies that the corresponding probability density is given by
$$p_\eta(x) = \frac{1}{b-a}, \tag{6.14}$$
so that if we indicate by μ a random variable uniformly distributed in the range $(0, 1)$, then η must satisfy the following equation:
$$\mu = \int_a^\eta p_\eta(x)\,dx = \int_a^\eta \frac{dx}{b-a}. \tag{6.15}$$
As a consequence, $\eta = a + \mu(b-a)$ and its expected value is given by $\langle \eta \rangle = (a+b)/2$.

6.2.4 Random Variable with Poisson Distribution

The Poisson distribution is another very important distribution for Monte Carlo simulations of electron-beam interactions with solid targets. This distribution is defined by the following probability density:
$$p_\chi(x) = \frac{1}{\lambda}\exp\left(-\frac{x}{\lambda}\right), \tag{6.16}$$
where λ equals a constant (whose meaning will be explained later). A random variable χ defined in the interval $(0, \infty)$, whose probability density is the Poisson distribution, is obtained by the solution of the equation
$$\mu = \int_0^\chi \frac{1}{\lambda}\exp\left(-\frac{x}{\lambda}\right)dx, \tag{6.17}$$

where μ indicates a random variable uniformly distributed in the range (0, 1). As a consequence, we can conclude that

$$\chi = -\lambda \ln(1-\mu). \tag{6.18}$$

As the distribution of $1-\mu$ equals that of μ, we can write

$$\chi = -\lambda \ln(\mu). \tag{6.19}$$

It is therefore possible to see that the constant λ is the expected value of χ:

$$\begin{aligned}\langle \chi \rangle &= \int_0^\infty x p_\chi(x)\,dx = \int_0^\infty \frac{x}{\lambda} \exp\left(-\frac{x}{\lambda}\right) dx \\ &= [-x \exp(-x/\lambda)]\big|_0^\infty + \int_0^\infty \exp(-x/\lambda)\,dx \\ &= -\lambda \exp(-x/\lambda)\big|_0^\infty = \lambda. \end{aligned} \tag{6.20}$$

6.2.5 Pseudo-Random-Number Generators

If we are able to generate random numbers uniformly distributed in the range (0, 1), it is possible, as we have previously shown, to generate random numbers distributed with any given probability density in any given interval. In other words, we can say that the uniformly distributed random variable underlies the generation of any other random number. As people generally use computer procedures that generate sequences of pseudo-random numbers uniformly distributed in the interval (0, 1), the problem arises of checking the uniformity of a given pseudo-random-number generator. A simple way to check the uniformity of a given distribution is to simulate the value of $\pi = 3.14\ldots$. This can be performed by generating numerous pairs of pseudo-random numbers distributed in the range $(-1, 1)$. If μ is a sequence of pseudo-random numbers distributed in the interval (0, 1) then $2\mu - 1$ is a sequence of pseudo-random numbers distributed in the interval $(-1, 1)$. If the number of pairs is statistically significant and the distribution is uniform, then the ratio between the number of pairs that lie in the unit circle and the total number of generated pairs should approach the value $\pi/4$. In other words, a comparison between the Monte Carlo-calculated area of the unit circle and its exact value can be used as a method to check the level of uniformity of a pseudo-random-number generator.

6.3 A Simple Monte Carlo Scheme

We can introduce a simple, reliable procedure, whereby a flux of electrons with energy E_0 impinges on a homogeneous target in the $+z$ direction. The electrons of the beam are elastically scattered by the atoms of the target, and lose their energy in inelastic collisions with the atomic electrons and through

6 Monte Carlo Simulations

plasmon excitations. For the simple procedure we are now describing, we assume that all the energy lost by the electrons of the incident beam can be described by utilising a continuous-energy-loss approximation. Electrons are thus assumed to lose energy continuously inside the solid, and the energy loss processes are all incorporated into the stopping power. The statistical fluctuations of the energy loss are therefore disregarded in this case.

The elastic scattering cross-sections are calculated assuming that the first Born approximation is valid throughout the electron travel in the solid target. The systematic errors introduced by this approximation are expected to be more important for the lowest primary energies.

We adopt spherical coordinates (r, θ, ϕ), and the path-length distribution is assumed to follow Poisson statistics, with a mean value equal to the electron elastic mean free path. The step length Δs is then given by

$$\Delta s = -\lambda_{el} \ln(\mu_1), \tag{6.21}$$

where μ_1 is a random number uniformly distributed in the range (0, 1) while λ_{el} represents the elastic mean free path. This is given by

$$\lambda_{el} = \frac{1}{N\sigma_{el}}, \tag{6.22}$$

where N is the number of atoms per unit of volume in the target and σ_{el} is the total elastic scattering cross-section

$$\sigma_{el} = 2\pi \int_0^\pi \frac{d\sigma}{d\Omega} \sin\theta \, d\theta. \tag{6.23}$$

For a Wentzel-like potential $V(r) = -(Ze^2/r)\exp(-r/a)$, in the first Born approximation, the total elastic scattering cross-section is given by

$$\sigma_{el} = \frac{\pi Z^2 e^4}{E^2} \frac{1}{\alpha(2+\alpha)}, \tag{6.24}$$

so that the elastic mean free path is

$$\lambda_{el} = \frac{\alpha(2+\alpha)E^2}{N\pi e^4 Z^2}, \tag{6.25}$$

where

$$\alpha = \frac{m e^4 \pi^2}{h^2} \frac{Z^{2/3}}{E}. \tag{6.26}$$

The polar scattering angle θ after an elastic collision is selected assuming that the probability $P(\theta)$ of elastic scattering into an angular range from 0 to θ is a random number μ_2, uniformly distributed in the range (0, 1). In other words, the integrated probability for scattering in the angular range from 0 to θ equals μ_2:

$$\mu_2 = P(\theta) = \frac{2\pi \int_0^\theta (d\sigma/d\Omega) \sin\vartheta \, d\vartheta}{2\pi \int_0^\pi (d\sigma/d\Omega) \sin\vartheta \, d\vartheta}. \tag{6.27}$$

From the equation
$$P(\theta) = \frac{(1+\alpha/2)(1-\cos\theta)}{1-\cos\theta+\alpha}, \tag{6.28}$$
it follows that
$$\cos\theta = 1 - \frac{2\alpha\mu_2}{2+\alpha-2\mu_2}. \tag{6.29}$$

The azimuthal angle ϕ can take on any value in the range $(0, 2\pi)$ and is selected by a random number η uniformly distributed in the interval $(0, 2\pi)$.

Note that both the polar scattering angle θ and the azimuthal angle ϕ are calculated relative to the last direction in which the electron was moving before the collision. The angle θ'_z at which the electron is moving after the last deflection, relative to the z direction, is given by

$$\cos\theta'_z = \cos\theta_z \cos\theta + \sin\theta_z \sin\theta \cos\phi, \tag{6.30}$$

where θ_z is the angle relative to the z direction before impact. The motion Δz along the z direction can then be obtained from

$$\Delta z = \Delta s \cos\theta'_z. \tag{6.31}$$

The new angle θ'_z is the incident angle θ_z for the next path length. Electrons are followed into the solid target until their energy becomes lower than the mean ionisation energy J (in eV, $J = 13.6Z$ for $Z \leq 10$; otherwise $J = (9.76 + 58.8Z^{-1.19})Z$) or until they emerge from the surface (backscattered particles) or from the back of the target (transmitted particles). In such a way, we avoid taking into account very low energies, for which the elastic-scattering calculation requires the partial-wave expansion method. This procedure is quite accurate if the primary energy of the impinging electron beam is higher than ~ 10 keV.

For these primary energies, the energy loss can be safely calculated using the Bethe formula [1],

$$-\frac{dE}{ds} = \frac{2\pi e^4 NZ}{E} \ln\left(\frac{1.166E}{J}\right), \tag{6.32}$$

or with semi-empirical expressions, such as the following,

$$-\frac{dE}{ds} = \frac{K_e N Z^{8/9}}{E^{2/3}}, \tag{6.33}$$

proposed in 1972 by Kanaya and Okayama (with $K_e = 360$ eV$^{5/3}$ Å^2) [2]. This last formula allows one to analytically evaluate the maximum range of penetration, whereby

$$R = \int_{E_0}^0 \frac{dE}{dE/ds} = \frac{3E_0^{5/3}}{5K_e N Z^{8/9}}. \tag{6.34}$$

Once the segment of the trajectory Δs has been calculated, the energy loss ΔE can be approximated by

$$\Delta E = \int_s^{s+\Delta s} \frac{dE}{ds} ds \approx \frac{dE}{ds} \Delta s \,. \tag{6.35}$$

Note that, with this approach, we neglect the energy loss straggling, i.e. the statistical fluctuations of the energy losses.

So, the simple Monte Carlo simulation described here can be safely utilised for primary energies higher than 10 keV in the calculation of the backscattering coefficient, for example [3, 4]: the backscattering coefficient is calculated by counting the number of electrons that emerge from the surface of a bulk target and dividing it by the total number of simulated trajectories (which should be more than 10 000 in order to obtain statistically significant values of the backscattering coefficient).

6.4 A More Sophisticated Simulation

The Monte Carlo scheme previously described can also be utilised for energies lower than 10 keV. To do this, it is necessary that the cross-sections utilised to describe the elastic scattering are calculated with the partial-wave expansion method, and the stopping power and the inelastic mean free path are computed utilising optical data and the dielectric function. With the previously described scheme, it is possible to simulate the interaction of electrons of low to medium primary energy (1–10 keV) and to extend each trajectory until the electron energy becomes lower than ∼50 eV, with these cross-sections and stopping powers.

The following analytical expression can be utilised, concerning the use of the relativistic partial-wave expansion method, to expedite the calculations and to approximate the differential elastic scattering cross-section [5, 6, 7]:

$$\frac{d\sigma}{d\Omega} = \Phi(Z,E) \frac{1}{(1 - \cos\theta + \Upsilon)^2} \,. \tag{6.36}$$

Note that, with $\Phi(Z,E) = Z^2 e^4/4E^2$ and $\Upsilon(Z,E) = 0$, this equation becomes identical to the classical Rutherford formula (3.38), while with $\Phi(Z,E) = Z^2 e^4/4E^2$ and $\Upsilon(Z,E) = \alpha(Z,E) = (me^4\pi^2/h^2)(Z^{2/3}/E)$, it becomes the screened Rutherford formula (corresponding to a Wentzel-like potential) (3.36). So, as (6.36) is identical to the equation we used in the previously described Monte Carlo scheme, we can use the same formal structure as in that procedure. However, it is necessary that $\Phi(Z,E)$ and $\Upsilon(Z,E)$ are computed in order to obtain, by using (6.36), the same values of the total cross-section σ_{el} and of the transport cross-section σ_{tr} as calculated by utilising the quantum-relativistic partial-wave expansion method. [8, 9, 10, 11]. Note that these quantities are defined, respectively, by the following:

$$\sigma_{el} = \int \frac{d\sigma}{d\Omega} d\Omega, \tag{6.37}$$

$$\sigma_{tr} = \int (1 - \cos\theta) \frac{d\sigma}{d\Omega} d\Omega. \tag{6.38}$$

Therefore, (6.36) allows one to sample the polar scattering angle with a closed formula similar to that used in the previous Monte Carlo scheme:

$$\cos\theta = 1 - \frac{2\Upsilon P(\theta)}{2 + \Upsilon - 2P(\theta)}. \tag{6.39}$$

Let us now show how to calculate $\Phi(Z,E)$ and $\Upsilon(Z,E)$ by use of our knowledge of the total elastic scattering cross-section σ_{el} and of the transport elastic scattering cross-section σ_{tr} mentioned above. Note that from (6.37) it follows that

$$\sigma_{el} = \Phi(Z,E) \frac{4\pi}{\Upsilon(2+\Upsilon)}, \tag{6.40}$$

so that the differential elastic scattering cross-section can be rewritten as

$$\frac{d\sigma}{d\Omega} = \frac{\sigma_{el}}{4\pi} \frac{\Upsilon(\Upsilon+2)}{(1-\cos\theta+\Upsilon)^2}. \tag{6.41}$$

Using (6.38) and (6.41), we obtain, concerning the transport cross-section,

$$\sigma_{tr} = \sigma_{el} \left[\frac{\Upsilon(\Upsilon+2)}{2} \ln\left(\frac{\Upsilon+2}{\Upsilon}\right) - \Upsilon \right]. \tag{6.42}$$

Let us calculate now the ratio Ξ between the transport elastic scattering cross-section and the total elastic scattering cross-section:

$$\Xi = \frac{\sigma_{tr}}{\sigma_{el}} = \Upsilon \left[\frac{\Upsilon+2}{2} \ln\left(\frac{\Upsilon+2}{\Upsilon}\right) - 1 \right]. \tag{6.43}$$

The values of σ_{tr} and of σ_{el} have been numerically calculated, so that it is possible to find Ξ as a function of Z and E. The values of Υ as a function of Z and of the electron energy E can be subsequently computed, by use of our knowledge of Ξ, by utilising a bisection algorithm. Once the values of $\Upsilon(Z,E)$ have been stored in a file, it is possible to calculate the value of Υ corresponding to the particle energy at every step of the electron trajectory during the simulation. The sampling of the scattering angle θ can be easily performed by inserting the value of Υ (found using an interpolation algorithm such as a cubic spline) into (6.39) and selecting a random number uniformly distributed in the range (0, 1) for $P(\theta)$.

6.4.1 Surface Films

The interface between the film and the substrate must be taken into account for surface films. After crossing the interface, the change in the scattering probabilities per unit length in passing from the film to the substrate and vice versa has to be considered. Let us denote by p_1 and p_2 the scattering probabilities per unit length for the two materials, where p_1 refers to the material in which the last elastic collision occurred and p_2 to the other material. d indicates the distance along the scattering direction between the initial scattering event and the interface. If rnd is a random number uniformly distributed in the range (0, 1), the step-length Δl is given by

$$\Delta l = \begin{cases} (1/p_1)[-\ln(1 - rnd)] \\ \quad \text{for } 0 \leq rnd < 1 - \exp(-p_1 d) \\ d + (1/p_2)[-\ln(1 - rnd) - p_1 d] \\ \quad \text{for } 1 - \exp(-p_1 d) \leq rnd \leq 1 \end{cases} . \tag{6.44}$$

If $p_1 = p_2$ (i.e. if there is no interface), the interface is only an imaginary boundary at a depth z below the surface and, for each value of rnd, $\Delta l = \lambda_1[-\ln(1-rnd)]$. If d is very large with respect to $1/p_1$, then $1-\exp(-p_1 d) \approx 1$ and, as a consequence, $\Delta l \approx (1/p_1)[-\ln(1-rnd)] = \lambda_1[-\ln(1-rnd)]$. In other words, when the particle is far away from the interface, its behaviour is practically the same as that described by (6.21). Note that, in general, if $rnd < 1 - \exp(-p_1 d)$ then the next elastic collision will occur in the same material and the interface will not be crossed. On the other hand, when $rnd \geq 1 - \exp(-p_1 d)$, then the next elastic collision will be in the other material because the interface will be crossed. In this case, the particle will travel a distance d along the scattering direction in the material where the last collision has occurred according to the scattering probability for that material and, after crossing the interface, it will travel a distance $\Delta l - d$ along the scattering direction according to the scattering probability of the other material. The energy loss is approximated by utilising the same random number rnd used to calculate Δl:

$$\Delta E = \begin{cases} (dE/dl)_1(1/p_1)[-\ln(1 - rnd)] \\ \quad \text{for } 0 \leq rnd < 1 - \exp(-p_1 d) \\ (dE/dl)_1 d + (dE/dl)_2(1/p_2)[-\ln(1 - rnd) - p_1 d] \\ \quad \text{for } 1 - \exp(-p_1 d) \leq rnd \leq 1 \end{cases} , \tag{6.45}$$

where $(-dE/dl)_1$ is the stopping power in the material where the last elastic collision occurred and $(-dE/dl)_2$ the stopping power in the other material. Note that this approach is correct only if the film thickness is greater than the electron elastic mean free path, which means that the (6.44) and (6.45) cannot be used when the film thickness is smaller than the electron elastic mean free path.

6.5 Another Monte Carlo Scheme

We shall now briefly describe another Monte Carlo scheme based on the classical-trajectory approximation. This scheme distinguishes the elastic and the inelastic collisions during electron travel inside the solid. The step length Δs is given by

$$\Delta s = -\lambda_t \ln(\mu_1), \tag{6.46}$$

where μ_1 is a random number uniformly distributed in the range (0, 1) and λ_t is the total mean free path:

$$\lambda_t = \frac{1}{\lambda_{inel}^{-1} + \lambda_{el}^{-1}}. \tag{6.47}$$

Before each collision, a random number uniformly distributed in the range (0, 1) is generated and compared with the probability of inelastic scattering p_{inel}. The probability of inelastic scattering is given by

$$p_{inel} = \lambda_{inel}^{-1}/(\lambda_{inel}^{-1} + \lambda_{el}^{-1}), \tag{6.48}$$

while, of course, that of elastic scattering is $p_{el} = 1 - p_{inel}$. If the random number generated is less than or equal to the probability of inelastic scattering, then the collision will be inelastic; otherwise, it will be elastic.

If the collision is elastic, then the calculation of the polar scattering angle proceeds as in the previously described Monte Carlo scheme. The azimuthal angle is calculated as a random number uniformly distributed in the range $(0, 2\pi)$.

If, on the other hand, the collision is inelastic, then the energy loss is computed by utilising a random number μ_2, uniformly distributed in the range (0, 1), so that

$$\mu_2 = \frac{\int_0^{\Delta E}[d\lambda_{inel}^{-1}/d(\Delta E')]d(\Delta E')}{\int_0^{E}[d\lambda_{inel}^{-1}/d(\Delta E')]d(\Delta E')}, \tag{6.49}$$

where ΔE is the energy transferred to a secondary electron. The secondary cascade may then be followed.

6.5.1 Angular Deflection in Electron–Electron Collisions

Note that, in the *elastic electron–electron* collisions corresponding to *inelastic electron–atom* scattering, the electrons not only lose energy but also suffer angular deflection.

To evaluate the angular deflections suffered by the electrons when the electron–atom collision is inelastic, we shall introduce the classical binary-collision model, which is sufficiently accurate for many practical purposes. In other words, we shall treat the scattering problem of the angular deflection in an *inelastic electron–atom* collision by studying an *elastic electron–electron* classical collision.

Let us consider an electron travelling along the z direction, and suppose that a free electron is at rest at a distance r. Let us indicate by v_0 the velocity of the incident electron (the primary electron) before impact, by v its velocity after impact, by u the velocity after impact of the electron initially at rest (the secondary electron), and by θ and ϑ the scattering angles of the primary and of the secondary electron, respectively, after impact.

Energy and momentum conservation imply that

$$v_0^2 = v^2 + u^2 , \tag{6.50}$$

$$v_0 = v\cos\theta + u\cos\vartheta , \tag{6.51}$$

$$v\sin\theta - u\sin\vartheta = 0 . \tag{6.52}$$

Therefore,

$$v^2 \cos^2\theta = (v_0 - u\cos\vartheta)^2 , \tag{6.53}$$

$$v^2 \sin^2\theta = u^2 \sin^2\vartheta \tag{6.54}$$

and, as a consequence,

$$2u^2 = 2uv_0 \cos\vartheta . \tag{6.55}$$

Then,

$$u = v_0 \cos\vartheta , \tag{6.56}$$

$$v^2 = v_0^2(1 - \cos^2\vartheta) = v_0^2 \sin^2\vartheta . \tag{6.57}$$

Moreover,

$$\sin^2\theta = \frac{u^2}{v^2} \sin^2\vartheta = \frac{v_0^2 \cos^2\vartheta}{v_0^2 \sin^2\vartheta} \sin^2\vartheta , \tag{6.58}$$

so that

$$\sin^2\theta = \cos^2\vartheta . \tag{6.59}$$

The energy of the primary electron is $E = mv_0^2/2$. Let us indicate by ΔE the energy lost by the primary electron and transferred to the secondary electron:

$$\Delta E = \frac{1}{2}mv_0^2 - \frac{1}{2}mv^2 = \frac{1}{2}mu^2 . \tag{6.60}$$

Straightforward algebraic manipulations allow us to conclude that

$$\frac{\Delta E}{E} = \cos^2\vartheta = \sin^2\theta . \tag{6.61}$$

6.5.2 Secondary Electrons

The incorporation of secondary electrons requires a knowledge of their initial angles and energies. If θ and ϕ are respectively the scattering and the azimuthal angle of the primary electron, using the classical binary-collision theory, the scattering angle ϑ of the secondary electron should satisfy the relationship

$$\cos\vartheta = \sin\theta \,, \tag{6.62}$$

and the azimuthal angle φ of the secondary electron is given by

$$\varphi = \pi + \phi \,. \tag{6.63}$$

The last two equations are used to represent the ionisation processes. Since slow secondary electrons are generated with spherical symmetry, in the case of the Fermi sea excitation we can assume a random direction of the secondary electrons:

$$\vartheta = \pi\mu_3 \,, \tag{6.64}$$

$$\varphi = 2\pi\mu_4 \,. \tag{6.65}$$

Here μ_3 and μ_4 are random numbers uniformly distributed in the range (0, 1). The primary-particle energy loss ΔE is transferred to the secondary electron. The electron that has less energy is generally considered as the secondary.

The surface energy barrier influences the energy distribution of the low-energy electrons, so that secondary electrons cannot emerge from the surface if they have an angle θ with respect to the normal higher than

$$\theta_c = \cos^{-1}\sqrt{\frac{E_f + U}{E + E_f + U}}\,, \tag{6.66}$$

where E_f is the Fermi energy and U the work function.

6.6 Comparing Theory and Experimental Data

In the previous chapters, we showed that, for scattering angles higher than 5°, the accuracy of the calculation of the differential elastic scattering cross-sections presented was 1–2%, while that for the total elastic scattering cross-section was of the order of 5–6%. On the other hand, the total cross-section does not feature strongly in multiple-scattering processes: a systematic study of the first transport elastic scattering cross-section has shown that the accuracy of the present approach is of the order of 2%. Taking into account the inaccuracies in (i) the evaluation of the elastic and inelastic cross-sections, (ii) the approximations introduced by using cubic spline interpolations of tabulated data, (iii) the stopping power used to calculate the energy losses and (iv) the statistical uncertainty of the Monte Carlo procedure, we are confident in concluding that our simulations give results with an accuracy of 5–15% for electron and positron energies higher than ~100 eV. In the case of lower energies, we may anticipate greater inaccuracies because the evaluation of the cross-sections is less accurate. In particular, in this low-energy regime (energies lower than 100–200 eV), the accuracy of the simulation should be evaluated by directly comparing the results with the experimental data available, as it is well known that theoretical evaluation of the relevant cross-sections is still an open problem.

In Fig. 6.1, we compare the Monte Carlo calculation results for the fractions of electrons absorbed by and transmitted through unsupported thin films of Cu of different thicknesses with the experimental data of Cosslett and Thomas [12]. In Figs. 6.2, 6.3, the electron backscattering coefficients for 5 keV and 10 keV electrons, respectively, impinging on various targets, are presented and compared with experimental data. The agreement between the Monte Carlo results and the experimental data considered here is satisfactory, with an accuracy of the order of 1–10%.

The depth distribution of 3 keV electrons absorbed in a 100 Å thin film of gold deposited on a silicon dioxide substrate is shown in Fig. 6.4, as calculated with the Monte Carlo code.

The fraction of electrons backscattered by a supported thin film $\zeta_B(z)$ as a function of the film thickness z can be directly calculated by a Monte Carlo simulation. On the other hand, by using Monte Carlo data concerning the fraction $\eta_B(z)$ for unsupported films of thickness z and the backscattering coefficients r_t and r_s of the materials constituting the film and the substrate, (5.67) can also be used to calculate $\zeta_B(z)$. A comparison between the values of $\zeta_B(z)$ obtained by these two different procedures allows us to further check

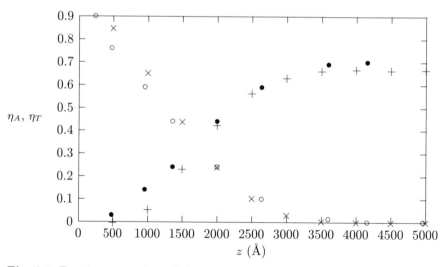

Fig. 6.1. Fractions η_A and η_T of electrons absorbed and transmitted, respectively, for unsupported thin films of Cu as a function of the film thickness z. The primary energy is 10 keV. •: η_A, Cosslett and Thomas experimental data [12]. +: η_A, Monte Carlo data [13]. ○: η_T, Cosslett and Thomas experimental data [12]. ×: η_T, Monte Carlo data [13]

6.6 Comparing Theory and Experimental Data   83

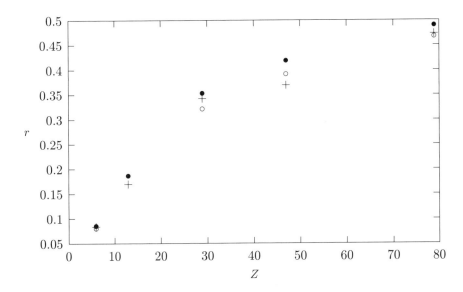

Fig. 6.2. Backscattering coefficient r as a function of the atomic number Z for streams of 5 keV electrons irradiating targets in the $+z$ direction. +: Monte Carlo data [13]. •: Bishop experimental data [14]. ○: Hunger and Küchler experimental data [15]

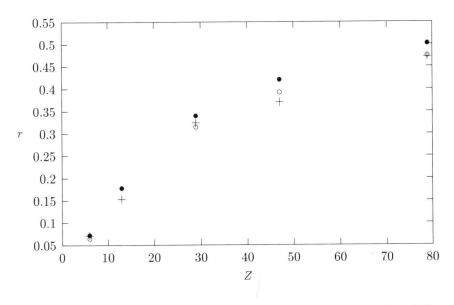

Fig. 6.3. Backscattering coefficient r as a function of the atomic number Z for streams of 10 keV electrons irradiating targets in the $+z$ direction. +: Monte Carlo data [13]. •: Bishop experimental data [14]. ○: Hunger and Küchler experimental data [15]

84   6 Monte Carlo Simulations

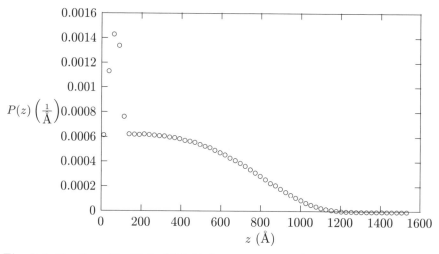

Fig. 6.4. Depth profile $P(z)$ of 3 keV electrons penetrating into a thin film of Au deposited on a SiO_2 substrate. Integration of $P(z)$ from $z = 0$ to $z = R$ gives the absorption coefficient $1 - r$. The Au film thickness is 100 Å. ○: Monte Carlo data

the accuracy of the multiple reflection method for studying electron backscattering from supported thin films. Such a comparison is shown in Figs. 6.5 and 6.6, where we show the result of simulating 5 and 10 keV electron penetration in gold thin films deposited on carbon substrates. The agreement is satisfactory: the differences between the direct Monte Carlo computations and the results obtained using (5.67) are always smaller than ∼10–15% of the Monte Carlo data. Also, the general trend as a function of the thickness is what one expects on the basis of physical considerations.

Note that inelastic effects are treated phenomenologically in the Monte Carlo approach: they are included in the analytical approach with exactly the same level of accuracy as in the Monte Carlo because the values of $\eta_B(z)$, r_t and r_s utilised in (5.67) are the same as those used in the code. However, even though we have used values of $\eta_B(z)$, r_t and r_s calculated by the Monte Carlo method in (5.67), note that the directly calculated Monte Carlo values of $\zeta_B(z)$ were obtained through independent computations.

We present our results concerning the stopping profiles of electrons implanted in silicon dioxide and of positrons implanted in copper, respectively, in Figs. 6.7 and 6.8. The primary energies considered were 3.0, 5.0 and 10.0 keV. The stopping profiles have been given as $RP(z, E_0)$, where R is the range and z is the depth inside the solid measured from the surface. Note that, for any given, fixed primary energy E_0, integration of $P(z, E_0)$ from $z = 0$ to $z = R$ gives the absorption coefficient. Each curve of implantation was obtained by simulating 10^6 electron or positron trajectories. We also performed similar calculations for carbon, silver and gold. Note that it is possible to calculate the backscattering coefficient r and the range R as

6.6 Comparing Theory and Experimental Data   85

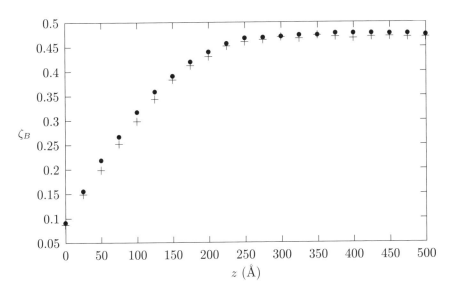

Fig. 6.5. Electron backscattering ratio ζ_B of Au surface films deposited on C bulk substrates vs surface film thickness z. The data presented concern 5 keV electron beams irradiating targets in the $+z$ direction. +: theoretical calculation based on the multiple reflection method (5.67) [13]. •: Monte Carlo data [13]

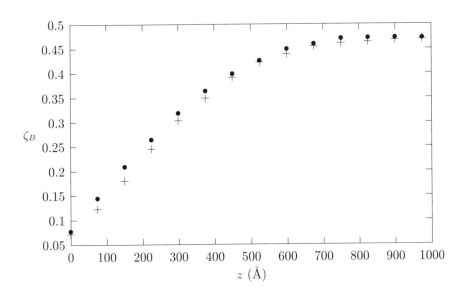

Fig. 6.6. Electron backscattering ratio ζ_B of Au surface films deposited on C bulk substrates vs surface film thickness z. The data presented concern 10 keV electron beams irradiating targets in the $+z$ direction. +: theoretical calculation based on the multiple reflection method (5.67) [13]. •: Monte Carlo data [13]

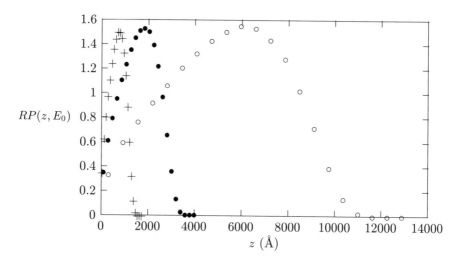

Fig. 6.7. Stopping profiles $RP(z, E_0)$ vs z (where R is the range, $P(z, E_0)$ the depth distribution, z the depth inside the solid measured from the surface and E_0 the primary energy) for electrons in SiO_2. For any given primary energy, integration of $P(z, E_0)$ from $z = 0$ to $z = R$ gives the absorption coefficient $1 - r$. The data presented concern Monte Carlo simulations of electron beams irradiating targets in the $+z$ direction [7]. +: 3 keV. •: 5 keV. ○: 10 keV

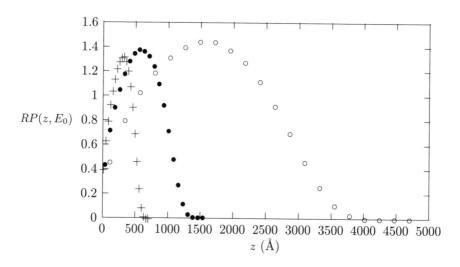

Fig. 6.8. Stopping profiles $RP(z, E_0)$ vs z (where R is the range, $P(z, E_0)$ the depth distribution, z the depth inside the solid measured from the surface and E_0 the primary energy) for positrons in Cu. For any given primary energy, integration of $P(z, E_0)$ from $z = 0$ to $z = R$ gives the absorption coefficient $1 - r$. The data presented concern Monte Carlo simulations of positron beams irradiating targets in the $+z$ direction [16]. +: 3 keV. •: 5 keV. ○: 10 keV

a function of the primary energy and of the atomic number, starting from the implantation profiles. The results reported compare excellently with both calculated and experimentally obtained data of other investigators. With regard to positrons, the calculated stopping profiles are an essential first step in solving the transport equation that is used in the application of the positron annihilation technique. The collisional processes of positron beams impinging on solids have recently received great attention, owing to the possibility of performing non-destructive investigations of point and extended defects of surfaces, interfaces and bulk materials by positron annihilation spectroscopy.

In conclusion, let us consider the problem of the secondary-electron emission from solids irradiated by a particle beam. It is an important problem, particularly in connection with analytical techniques which utilise secondary electrons to investigate chemical and compositional properties of solids in the near-surface layers, namely Auger electron spectroscopy and X-ray photoelectron spectroscopy. In general, the energy spectra of the electrons emitted are quite complicated because many features appear in such spectra, related to the different collisional processes involved before low-energy secondary-electron emission. As a consequence, a better understanding of the collisional events occuring in the surface layers before emission should allow a more general understanding of surface physics, including, for example, plasmon excitation.

When a particle beam (with energy exceeding some threshold value) impinges on a solid target, it stimulates the emission of secondary electrons through collisions with target atoms. On the other hand, a fraction of the particles of the primary beam is also ejected from the surface because some particles come back and emerge from the surface after a number of elastic and/or inelastic collisions with the target atoms. If the target is not a thin film (i.e. if there are no transmitted particles), the remaining primary particles are trapped in it. The ratio between the numbers of backscattered and total particles (backscattered + trapped) is generally called, for bulk targets, the backscattering coefficient. If the primary particles are electrons, then the spectrum of the secondary electrons is clearly contaminated by the contribution of the backscattered primary electrons. On the other hand, as recently noted by Overton and Coleman [17], the problem of distinguishing between true secondary electrons and backscattered electrons is absent if the secondary-electron emission is stimulated by a positron beam. The authors of [17] performed an interesting experimental study of the spectra of fast secondary electrons uncontaminated by backscattered electrons. In fact, analysis was carried out for secondary electrons produced by positron beams impinging on a copper target at a glancing angle and at a $35°$ incidence angle, for primary energies in the range from 50 eV to 2 keV [17]. Our results concerning the differential spectra of the energy distribution function of the secondary electrons emitted from a copper target above the low-energy cascade peak always show a clear linear trend, in agreement with both the

88   6 Monte Carlo Simulations

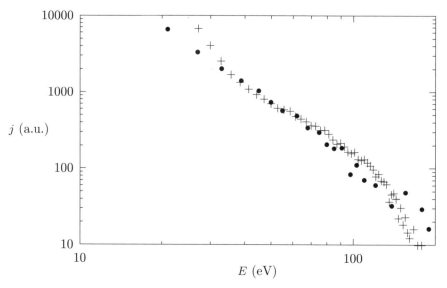

Fig. 6.9. Energy distribution of the secondary electrons emitted from Cu stimulated by a 300 eV positron beam. +: Monte Carlo data [18]. •: Overton and Coleman experimental data [17]

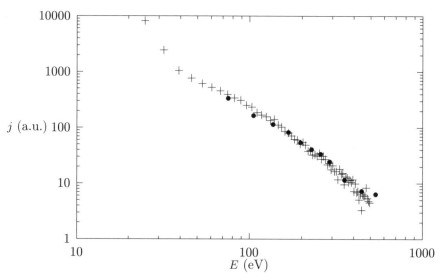

Fig. 6.10. Energy distribution of the secondary electrons emitted from Cu stimulated by a 1000 eV positron beam. +: Monte Carlo data [18]. •: Overton and Coleman experimental data [17]

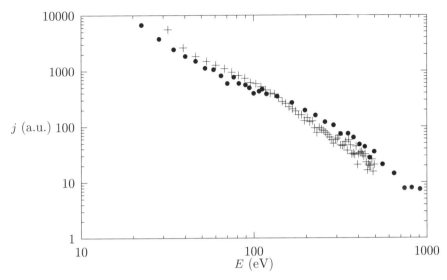

Fig. 6.11. Energy distribution of the secondary electrons emitted from Cu stimulated by a 2000 eV positron beam. ∘: Monte Carlo data [18]. •: Overton and Coleman experimental data [17]

Sickafus law [8, 9, 10] and Overton and Coleman's [17] experimental data (in the positron primary-energy range from 200 eV to 2 keV) when plotted on a log–log (base 10) scale [18] . The calculation of the coefficient m of the Sickafus law has been performed for the positron primary-energy range from 200 eV to 2 keV at different incidence angles without any significant dependence on the incidence angle being observed, in agreement with the experimental results. In Figs. 6.9–6.11 we present our calculated energy spectra of the secondary electrons emitted following irradiation of copper with 300, 1000 and 2000 eV positrons, respectively: the comparison with the experimental data of Overton and Coleman [17] demonstrates excellent agreement. The numerical results for m, as a function of the incident positron energy, together with the Overton and Coleman experimental data, are reported in Fig. 6.12. The incidence angle used in the calculation was 35° with respect to the surface of the sample. The results show excellent agreement for positron primary energies higher than ≈100 eV. However, when the positron primary energy is 100 eV or less, there is no agreement between theory and experiment. We attribute this discrepancy to the fact that the energy loss cannot be regarded as continuous in such a low-energy regime. In fact, to describe energy loss in such a low energy-regime, more appropriate calculations are necessary (quantum Monte Carlo method).

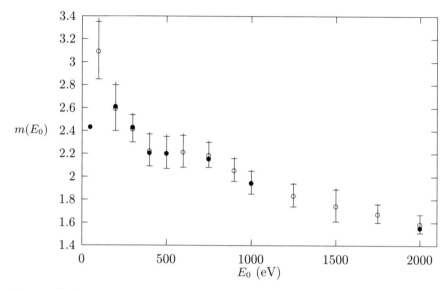

Fig. 6.12. Sickafus index m for secondary-electron emission from Cu stimulated by positrons, as a function of the positron primary energy E_0. +: Monte Carlo data [18], including numerical errors due to statistical sampling. •: Overton and Coleman experimental data [17]

References

1. H.A. Bethe, in *Handbuch der Physik* (Springer, Berlin, 1933), **24**, 519
2. K. Kanaya and S. Okayama, J. Phys. D **5**, 43 (1972)
3. M. Dapor, Phys. Rev. B **46**, 618 (1992)
4. M. Dapor, Appl. Surf. Sci. **70/71**, 327 (1993)
5. J. Baró, J. Sempau, J.M. Fernández-Varea, F. Salvat, Nucl. Instrum. Methods Phys. Res. B **84**, 465 (1994)
6. J. Baró, J. Sempau, J.M. Fernández-Varea, F. Salvat, Nucl. Instrum. Methods Phys. Res. B **100**, 31 (1995)
7. A. Miotello, M. Dapor, Phys. Rev. B **56**, 2241 (1997)
8. M. Dapor, Nucl. Instrum. Methods Phys. Res. B **95**, 470 (1995); Nucl. Instrum. Methods Phys. Res. B **108**, 363 (1996)
9. M. Dapor, J. Appl. Phys. **77**, 2840 (1995)
10. M. Dapor, J. Appl. Phys. **79**, 8406 (1996)
11. M. Dapor, Scanning Microsc. **9**, 939 (1995)
12. V.E. Cosslett, R.N. Thomas, Br. J. Appl. Phys. **16**, 779 (1965)
13. M. Dapor, Eur. Phys. J.: Appl. Phys. **18**, 155 (2002)
14. H.E. Bishop, *Proc. 4ème Congrès International d'Optique des Rayons X et de Microanalyse* (Hermann, Paris, 1967), pp. 153–158
15. H.-J. Hunger, L. Küchler, Phys. Status Solidi A **56**, K45 (1979)
16. M. Dapor, A. Miotello, in *Advanced Monte Carlo for Radiation Physics, Particle Transport Simulation and Applications*, ed. by A. Kling, F. Barão, M. Nakagawa, L. Távora, P. Vaz, (Springer, Berlin, Heidelberg, 2001), pp. 43–47
17. N. Overton, P.G. Coleman, Phys. Rev. Lett. **79**, 305 (1997)
18. M. Dapor, D. Zari, A. Miotello, Phys. Rev. B **61**, 5979 (2000)

A Matrices and Operators

A.1 Representation of Linear Operators

It is always possible to represent a linear operator L as a matrix. The representation of the linear operator L on the basis represented by a complete set of eigenfunctions $\{u_n(\mathbf{r})\}$ is given by the matrix constituted by the following matrix elements:

$$L_{nm} = (u_n, Lu_m) = \int u_n^* L u_m \, d^3x \; . \tag{A.1}$$

If $L = L^\dagger$, then $L_{nm}^* = L_{mn}$. In detail,

$$(L^\dagger)_{nm} = (u_n, L^\dagger u_m) = (Lu_n, u_m) = (u_m, Lu_n)^* = L_{mn}^* \; . \tag{A.2}$$

If $\{u_n(\mathbf{r})\}$ is an orthonormal set of eigenfunctions of the Hilbert space, then

$$(u_n, u_m) = \delta_{nm} \; . \tag{A.3}$$

If $\{u_n(\mathbf{r})\}$ is a set of orthonormal eigenfunctions of the operator L, then the representation of L on the basis $\{u_n(\mathbf{r})\}$ is a diagonal matrix. This can be written as

$$Lu_n = \lambda_n u_n \; , \tag{A.4}$$

and, as a consequence,

$$L_{nm} = (u_n, Lu_m) = \lambda_n (u_n, u_m) = \lambda_n \delta_{nm} \; . \tag{A.5}$$

A.2 Matrix Transformations

Let us introduce two complete sets of eigenfunctions $\{u_n(\mathbf{r})\}$ and $\{u'_n(\mathbf{r})\}$. If $\varphi(\mathbf{r})$ is a function in a Hilbert space, then

$$\varphi(\mathbf{r}) = \sum_n c_n u_n(\mathbf{r}) = \sum_n c'_n u'_n(\mathbf{r}) \; , \tag{A.6}$$

where

$$c_n = (u_n, \varphi) \tag{A.7}$$

and
$$c'_n = (u'_n, \varphi) \,. \tag{A.8}$$
We can therefore write that
$$u'_n = \sum_m T_{mn} u_m \,. \tag{A.9}$$
The elements of the matrix T can be calculated from
$$T_{mn} = (u_m, u'_n) = \int d^3r \, u^*_m(\boldsymbol{r}) u'_n(\boldsymbol{r}) \,. \tag{A.10}$$
The matrix T is unitary, i.e. $TT^\dagger = T^\dagger T = I$ where I represents the unity matrix:
$$(TT^\dagger)_{mp} = \sum_n T_{mn} T^*_{pn} = \int\int d^3r \, d^3r' \, u^*_m(\boldsymbol{r}) u_p(\boldsymbol{r}') \delta(\boldsymbol{r}' - \boldsymbol{r}) \,, \tag{A.11}$$
where we have used the closure relation
$$\sum_n u'_n(\boldsymbol{r}) u'^*_n(\boldsymbol{r}') = \delta(\boldsymbol{r} - \boldsymbol{r}') \,. \tag{A.12}$$
Therefore,
$$(TT^\dagger)_{mp} = \int d^3r \, u^*_m(\boldsymbol{r}) u_p(\boldsymbol{r}) = (u_m, u_p) = \delta_{mp} \,, \tag{A.13}$$
or
$$TT^\dagger = I \,. \tag{A.14}$$
In a similar way, we can show that $T^\dagger T = I$.

Note that
$$c'_n = \sum_m T^*_{mn} c_m \,. \tag{A.15}$$
In detail,
$$c'_n = (u'_n, \varphi) = \sum_m T^*_{mn}(u_m, \varphi) = \sum_m T^*_{mn} c_m \,. \tag{A.16}$$

Let us indicate by A a Hermitian matrix expressed in the basis u_n, and by A' its representation in the basis u'_n. We can write that
$$A' = T^\dagger A T \,. \tag{A.17}$$
In detail,
$$A'_{nm} = (u'_n, A u'_m) = \sum_{jk} (T_{jn} u_j, A T_{km} u_k) \,. \tag{A.18}$$
Now
$$\sum_{jk} (T_{jn} u_j, A T_{km} u_k) = \sum_{jk} T^*_{jn}(u_j, A u_k) T_{km} = \sum_{jk} (T^\dagger)_{nj} A_{jk} T_{km} \,, \tag{A.19}$$

and therefore
$$A'_{nm} = \sum_{jk}(T^\dagger)_{nj}A_{jk}T_{km} \ . \tag{A.20}$$

Different representations of the same operators are related by unitary transformations. These transformations do not change the norm of vectors:
$$(\phi,\phi) = (\phi, T^\dagger T \phi) = (T\phi, T\phi) \ . \tag{A.21}$$

If A is Hermitian, then A' is also Hermitian:
$$(A')^\dagger = (T^\dagger A T)^\dagger = (T^\dagger)(T^\dagger A)^\dagger = (T^\dagger A^\dagger T^{\dagger\dagger}) = T^\dagger A T = A' \ . \tag{A.22}$$

The Hermitian matrix A can be diagonalised by a unitary transformation, and the diagonal elements are matrix eigenvalues, which are real and can be obtained from the following equation:
$$\det(A - \lambda I) = 0 \ . \tag{A.23}$$

A.3 Commuting Operators

Let us show that if two operators have a complete set of eigenfunctions in common, then they commute. Let u_i be a complete set of eigenfunctions that the operators A and B have in common. If a_i and b_i are the eigenvalues, $Au_i = a_i u_i$ and $Bu_i = b_i u_i$. As $[A,B]u_i = (a_i b_i - b_i a_i)u_i = 0$ and
$$[A,B]\varphi = [A,B]\sum_i c_i u_i \ , \tag{A.24}$$
then as a consequence,
$$[A,B]\varphi = 0 \tag{A.25}$$
for every function in the Hilbert space. This means that $[A,B] = 0$.

Let us now assume that the operators A and B commute. Let be a an eigenvalue of A and N the dimension of the eigenspace corresponding to a, i.e.
$$Au_i = au_i \ , \tag{A.26}$$
for each $i = 1, \ldots, N$. As $ABu_i = BAu_i$, then $A(Bu_i) = a(Bu_i)$, so that the vectors Bu_i belong to the eigenspace corresponding to a, or
$$Bu_j = \sum_{i=1}^{N} c_{ji} u_i \ . \tag{A.27}$$

When $N = 1$, we can write
$$Au_1 = au_1 \tag{A.28}$$

and
$$Bu_1 = c_{11}u_1 \,, \tag{A.29}$$
so that, as a consequence, A and B have a complete set of eigenfunctions in common, represented by the single eigenvector u_1.

In general, if
$$c_{ji} = (u_i, Bu_j) \tag{A.30}$$
and B is Hermitian, then
$$c_{ji}^* = (Bu_j, u_i) = (u_j, Bu_i) = c_{ij} \,, \tag{A.31}$$
so that there is a unitary transformation s such that the matrix d defined by $s^\dagger c s = d$ is diagonal. If d_j are the diagonal elements of d, they are real because $d^\dagger = d$. So we can write that
$$\sum_{j=1}^N c_{ij}s_{jk} = \sum_{j=1}^N s_{ij}d_j\delta_{jk} = s_{ik}d_k \tag{A.32}$$
and, as a consequence,
$$\sum_{j=1}^N s_{jk}^* c_{ji} = d_k s_{ik}^* \,. \tag{A.33}$$
Utilising (A.27) and (A.33), we can write
$$B\sum_{j=1}^N s_{jk}^* u_j = \sum_{j=1}^N s_{jk}^* Bu_j = \sum_{j=1}^N \sum_{i=1}^N s_{jk}^* c_{ji} u_i = d_k \sum_{i=1}^N s_{ik}^* u_i \,. \tag{A.34}$$
Defining
$$v_k = \sum s_{ik}^* u_i \,, \tag{A.35}$$
we can conclude that
$$Av_k = av_k \tag{A.36}$$
and
$$Bv_k = d_k v_k \,, \tag{A.37}$$
so that A and B have a basis of eigenfunctions ($\{v_k\}$) in common.

B The Dirac Notation

B.1 Ket and Bra Vectors

In the Dirac notation, we use the symbol $|u\rangle$ to represent vectors. These are called "ket vectors". Linear combinations of kets are kets. A set of kets is said to be linearly independent if none of the kets is a linear combination of the others. When a vector space has exactly N linearly independent vectors, it is said to have N dimensions.

The dual space is the space of the "bra vectors", which are indicated by the symbol $\langle v|$. The symbol $\langle u|$ indicates the bra vector corresponding to the ket vector $|u\rangle$. The relation between a ket vector and the corresponding bra vector is the same as that between a wave function and its complex conjugate. The scalar product is represented by the *bracket* $\langle|\rangle$ and possesses the following properties: if $|u\rangle$ and $|v\rangle$ are two kets, then their scalar product is the product between the bra vector $\langle u|$ and the ket vector $|v\rangle$, and $\langle u|v\rangle = \langle v|u\rangle^*$. The norm of $|u\rangle$ is real and non-negative and is given by the following:

$$\|u\|^2 = \langle u|u\rangle . \tag{B.1}$$

$\langle u|u\rangle$ is zero if and only if $u = 0$. Each linear correspondence between two kets $|u\rangle$ and $|v\rangle$ defines a linear operator A such that

$$|v\rangle = A|u\rangle . \tag{B.2}$$

Let us now consider a complete set of eigenvectors $|n\rangle$ such that

$$|v\rangle = \sum_n c_n |n\rangle = \sum_n |n\rangle\langle n|v\rangle . \tag{B.3}$$

The coefficients c_n of the expansion are given by

$$c_n = \langle n|v\rangle . \tag{B.4}$$

Consequently,

$$A|u\rangle = \sum_n |n\rangle\langle n|A|u\rangle . \tag{B.5}$$

The set $|n\rangle$ is said to be a complete set of orthonormal eigenvectors if both the orthonormality condition,

$$\langle n|m\rangle = (u_n, u_m) = \delta_{nm} , \tag{B.6}$$

and the closure relation,

$$\sum_n |n\rangle\langle n| = I \,, \tag{B.7}$$

hold.

Let us consider a generic vector $|\varphi\rangle$. This vector can be expanded in a linear combination of eigenvectors. This can be written as

$$\sum_n |n\rangle\langle n|\varphi\rangle = \left(\sum_n |n\rangle\langle n|\right)|\varphi\rangle = I|\varphi\rangle = |\varphi\rangle \,. \tag{B.8}$$

Let us calculate the matrix elements of a linear operator:

$$L = \sum_n |n\rangle\langle n|L \sum_m |m\rangle\langle m| - \sum_n \sum_m |n\rangle\langle m|\langle n|L|m\rangle \,. \tag{B.9}$$

As a consequence, we can write that

$$L_{nm} = \langle n|L|m\rangle \,. \tag{B.10}$$

Let us now define the adjoint operator of L, $L^\dagger$, by

$$\langle u|L|v\rangle^* = \langle v|L^\dagger|u\rangle \,. \tag{B.11}$$

If $|v\rangle = A|u\rangle$, then $\langle v| = \langle u|A^\dagger$. In detail,

$$\langle v| = \sum_n \langle n|A|u\rangle^* \langle n| = \sum_n \langle u|A^\dagger|n\rangle\langle n| = \langle u|A^\dagger \,. \tag{B.12}$$

A projector P is defined as a linear operator such that

$$PP = P \,. \tag{B.13}$$

The projector on the direction of the normalised ket $|u\rangle$ is

$$P_u = |u\rangle\langle u| \,. \tag{B.14}$$

If $|\phi\rangle$ is a ket vector, then

$$P_u|\phi\rangle = |u\rangle\langle u|\phi\rangle = (\langle u|\phi\rangle)|u\rangle \,. \tag{B.15}$$

Let us now show that P_u is a projector:

$$P_u P_u = |u\rangle\langle u|u\rangle\langle u| = |u\rangle\langle u| = P_u \,. \tag{B.16}$$

B.2 Continuous Spectrum

Let us consider the eigenvalue equation that defines the spectrum of the position operator x,

$$x\varphi_a(x) = a\varphi_a(x) \,, \tag{B.17}$$

with eigenfunctions given by

$$\varphi_a(x) = \delta(x-a) . \tag{B.18}$$

The spectrum of the positions of a particle is characterised by a continuous set of eigenfunctions. They are δ distributions, i.e. (intuitively) functions that are zero in all cases apart from when $x = a$, where they become ∞.

If we introduce the orthogonality relation for a continuous spectrum, the scalar product between two eigenfunctions belonging to two different eigenvalues a and b is given by

$$\langle a|b\rangle = (\varphi_a, \varphi_b) = \int dx\, \delta(x-a)\delta(x-b) = \delta(a-b) . \tag{B.19}$$

The closure relation for a continuous spectrum is given by

$$\int da\, \varphi_a(x_1)\varphi_a(x_2) = \int da\, \delta(x_1-a)\delta(x_2-a) = \delta(x_1-x_2) . \tag{B.20}$$

Let us consider a wave function $\varphi(x)$,

$$\varphi(x) = \int dx'\, \delta(x-x')\varphi(x') , \tag{B.21}$$

and apply the closure relation

$$\varphi(x) = \int dx' \int da\, \varphi_a(x)\varphi_a(x')\varphi(x') . \tag{B.22}$$

Then

$$\varphi(x) = \int da\, c_a \varphi_a(x) , \tag{B.23}$$

where

$$c_a = \int dx'\, \varphi_a(x')\varphi(x') = (\varphi_a, \varphi) = \langle a|\varphi\rangle = \varphi(a) . \tag{B.24}$$

The coefficients of the expansion are given by the wave equation and, as a consequence, $|\varphi(x)|^2$ is the probability density of the position operator x. Let us now calculate the matrix elements for the position x and the x component of the momentum $p = (\hbar/i)(\partial/\partial x)$:

$$x_{ab} = \langle a|x|b\rangle = (\varphi_a, x\varphi_b) = \int dx\, \delta(x-a) x \delta(x-b) = a\delta(a-b) \tag{B.25}$$

and

$$p_{ab} = \langle a|p|b\rangle = (\varphi_a, p\varphi_b) = \int dx\, \delta(x-a) \frac{\hbar}{i}\frac{\partial}{\partial x} \delta(x-b)$$

$$= \frac{\hbar}{i}\frac{\partial}{\partial a}\delta(a-b) . \tag{B.26}$$

B.3 The Schrödinger Equation in the Dirac Notation

The Schrödinger equation in the Dirac notation is given by the following:

$$i\hbar \frac{\partial}{\partial t}|\psi, t\rangle = H|\psi, t\rangle \ . \tag{B.27}$$

From (B.27), let us show that we can obtain the Schrödinger equation in the usual notation. Note that the wave function is the projection of $|\psi\rangle$ on $\langle x|$:

$$\langle x|\psi, t\rangle = \psi(x, t) \ . \tag{B.28}$$

From (B.27), we obtain the result that

$$i\hbar \langle x| \frac{\partial}{\partial t}|\psi, t\rangle = \langle x|H|\psi, t\rangle \tag{B.29}$$

and, taking the closure relation into account,

$$i\hbar \langle x| \frac{\partial}{\partial t}|\psi, t\rangle = I\langle x|H|\psi, t\rangle = \int d\xi \langle x|H|\xi\rangle \langle \xi|\psi, t\rangle \ . \tag{B.30}$$

$\langle x|$ is independent of time, and

$$\langle x|H|\xi\rangle = \langle x|\frac{p^2}{2m} + V|\xi\rangle = -\frac{\hbar^2}{2m}\frac{\partial^2}{\partial x^2}\delta(x-\xi) + V(x)\delta(x-\xi) \ . \tag{B.31}$$

Therefore,

$$i\hbar \frac{\partial}{\partial t}\psi(x,t) = \left[-\frac{\hbar^2}{2m}\frac{\partial^2}{\partial x^2} + V(x) \right]\psi(x,t) \ . \tag{B.32}$$

This is the Schrödinger equation in the usual notation.

C Special Functions

C.1 Legendre Polynomials and Associated Legendre Functions

The Legendre polynomial of degree l ($l = 0, 1, 2, \ldots, \infty$) is defined by the formula

$$P_l(u) = \frac{1}{2^l l!} \frac{d^l}{du^l} (u^2 - 1)^l . \tag{C.1}$$

It is a polynomial with l zeros in the range $(-1, +1)$ and with parity $(-1)^l$. The first five Legendre polynomials are

$$P_0 = 1 , \tag{C.2}$$

$$P_1 = u , \tag{C.3}$$

$$P_2 = \frac{1}{2}(3u^2 - 1) , \tag{C.4}$$

$$P_3 = \frac{1}{2}(5u^3 - 3u) , \tag{C.5}$$

$$P_4 = \frac{1}{8}(35u^4 - 30u^2 + 3) . \tag{C.6}$$

Let us now introduce the associated Legendre functions:

$$P_l^m(u) = (1 - u^2)^{(1/2)m} \frac{d^m}{du^m} P_l(u) . \tag{C.7}$$

The Legendre polynomials are the particular associated Legendre functions corresponding to $m = 0$:

$$P_l(u) = P_l^0(u) . \tag{C.8}$$

The associated Legendre functions satisfy the orthonormality relations

$$\int_{-1}^{+1} P_k^m P_l^m \, du = \frac{2}{2l+1} \frac{(l+m)!}{(l-m)!} \delta_{kl} , \tag{C.9}$$

and the differential equation

$$\left[(1-u^2)\frac{d^2}{du^2} - 2u\frac{d}{du} + l(l+1) - \frac{m^2}{1-u^2}\right]P_l^m = 0 \,. \tag{C.10}$$

The following recursion relations are very useful when using the Legendre polynomials:

$$(l-m+1)P_{l+1}^m(u) + (l+m)P_{l-1}^m(u) = (2l+1)uP_l^m(u) \,, \tag{C.11}$$

$$(1-u^2)\frac{d}{du}P_l^m(u) = (l+m)P_{l-1}^m(u) - luP_l^m(u) \,. \tag{C.12}$$

In order to find the recursion relations corresponding to the Legendre polynomials, we can impose $m = 0$ in the previous equations and obtain the following:

$$(l+1)P_{l+1}(u) + lP_{l-1}(u) = (2l+1)uP_l(u) \,, \tag{C.13}$$

$$(1-u^2)\frac{d}{du}P_l(u) = lP_{l-1}(u) - luP_l(u) \,. \tag{C.14}$$

The Legendre polynomials are eigenfunctions of the square of the orbital angular momentum:

$$\boldsymbol{L}^2 = -\hbar^2\left(\frac{\partial^2}{\partial\theta^2} + \cot\theta\frac{\partial}{\partial\theta} + \frac{1}{\sin^2\theta}\frac{\partial^2}{\partial\phi^2}\right) \,, \tag{C.15}$$

$$\boldsymbol{L}^2 P_l(\cos\theta) = \hbar^2 l(l+1)P_l(\cos\theta) \,. \tag{C.16}$$

C.2 Bessel Functions

Let us introduce the Bessel equation of order ν:

$$x^2\frac{d^2y}{dx^2} + x\frac{dy}{dx} + (x^2 - \nu^2)y = 0 \,. \tag{C.17}$$

The solution of this equation is a linear combination of the Bessel functions $J_{-\nu}$ and $J_{+\nu}$.

Let us now consider the Schrödinger equation of a particle in a constant potential V_0,

$$(\boldsymbol{\nabla}^2 + \boldsymbol{K}^2 - U_0)\Psi = 0 \,, \tag{C.18}$$

where $\boldsymbol{\nabla} \equiv (\partial/\partial x, \partial/\partial y, \partial/\partial z)$, $\boldsymbol{K} = \boldsymbol{p}/m$, $K^2 = 2mE/\hbar^2$ and $U_0 = 2mV_0/\hbar^2$. In order to proceed, let us expand the wave function in Legendre polynomials:

$$\Psi(r, \cos\theta) = \sum_{l=0}^{\infty} a_l \frac{y_l(r)}{r} P_l(\cos\theta) \,. \tag{C.19}$$

C.2 Bessel Functions

Taking into account (C.16), the Schrödinger equation becomes the following:

$$\sum_{l=0}^{\infty} \left[\frac{\partial^2}{\partial r^2} + \frac{2}{r}\frac{\partial}{\partial r} - \frac{l(l+1)}{r^2} + K^2 - U_0 \right] a_l \frac{y_l(r)}{r} P_l(\cos\theta) = 0 \ . \qquad (\text{C.20})$$

All the coefficients of the expansion must then satisfy the following differential equations:

$$\left[\frac{\partial^2}{\partial r^2} + \frac{2}{r}\frac{\partial}{\partial r} - \frac{l(l+1)}{r^2} + K^2 - U_0 \right] a_l \frac{y_l(r)}{r} = 0 \ . \qquad (\text{C.21})$$

If we define

$$k^2 = K^2 - U_0 \ , \qquad (\text{C.22})$$

we can write

$$\left[\frac{d^2}{dr^2} - \frac{l(l+1)}{r^2} + k^2 \right] y_l(r) = 0 \ . \qquad (\text{C.23})$$

Let us now introduce the variable $x \equiv kr$, so that the last equation can be rewritten as

$$\left[\frac{d^2}{dx^2} - \frac{l(l+1)}{x^2} + 1 \right] y_l(x) = 0 \ . \qquad (\text{C.24})$$

The *spherical Bessel function of order l* is defined by

$$j_l(x) = \sqrt{\frac{\pi}{2x}} J_{l+1/2}(x) \ . \qquad (\text{C.25})$$

Since $x^{1/2} J_{l+1/2}$ is a solution of (C.24), we can conclude that the function $kr j_l(kr)$ is a solution of (C.23).

In a similar way, once the *spherical Neumann function of order l* (also known as the *irregular spherical Bessel function of order l*) has been defined as

$$n_l(x) = (-1)^{l+1} \sqrt{\frac{\pi}{2x}} J_{-l-1/2}(x) \ , \qquad (\text{C.26})$$

it is possible to show that $kr n_l(kr)$ is a solution of (C.23).

The first three regular spherical Bessel functions are

$$j_0 = \frac{\sin x}{x} \ , \qquad (\text{C.27})$$

$$j_1 = \frac{\sin x}{x^2} - \frac{\cos x}{x} \ , \qquad (\text{C.28})$$

$$j_2 = \left(\frac{3}{x^3} - \frac{1}{x} \right) \sin x - \frac{3}{x^2} \cos x \ , \qquad (\text{C.29})$$

while the first three spherical Neumann functions are

$$n_0 = -\frac{\cos x}{x} \ , \qquad (\text{C.30})$$

102   C Special Functions

$$n_1 = -\frac{\cos x}{x^2} - \frac{\sin x}{x}, \qquad (C.31)$$

$$n_2 = \left(-\frac{3}{x^3} + \frac{1}{x}\right)\cos x - \frac{3}{x^2}\sin x. \qquad (C.32)$$

It can be proved that

$$j_l(x) \underset{x\to 0}{\sim} \frac{x^l}{1\cdot 3\cdot\ldots\cdot(2l+1)}, \qquad (C.33)$$

and that

$$n_l(x) \underset{x\to 0}{\sim} -\frac{1\cdot 3\cdot\ldots\cdot(2l-1)}{x^{l+1}}. \qquad (C.34)$$

The following equation holds:

$$j_l(0) = \delta_{l0}. \qquad (C.35)$$

The asymptotic behaviour of the Bessel and Neumann functions is described by the following equations:

$$j_l(x) \underset{x\to\infty}{\sim} \frac{\sin(x - l\pi/2)}{x}, \qquad (C.36)$$

$$n_l(x) \underset{x\to\infty}{\sim} -\frac{\cos(x - l\pi/2)}{x}. \qquad (C.37)$$

If we indicate by f_l any linear combination of the Bessel and Neumann functions ($f_l = aj_l + bn_l$, where a and b are arbitrary coefficients), we have

$$x f_{l-1} - (2l+1)f_l + x f_{l+1} = 0, \qquad (C.38)$$

$$x f_{l-1} - (l+1)f_l - x\frac{df_l}{dx} = 0. \qquad (C.39)$$

C.3 The Spherical Harmonics

The spherical harmonics Y_l^m are the eigenfunctions common to the operators $\boldsymbol{L}^2$ and L_z. With $l = 0, 1, \ldots$ and $m = -l, -l+1, \ldots, l-1, l$, we have

$$\boldsymbol{L}^2 Y_l^m = \hbar^2 l(l+1) Y_l^m \qquad (C.40)$$

and

$$L_z Y_l^m = \hbar m Y_l^m. \qquad (C.41)$$

C.3 The Spherical Harmonics

When $m \geq 0$, the spherical harmonics are given by the following:

$$Y_l^m(\theta, \phi) = (-1)^m \sqrt{\frac{(2l+1)(l-m)!}{4\pi(l+m)!}} P_l^m(\cos\theta) \exp(im\phi) ,\qquad (C.42)$$

whereas, when $m < 0$, the spherical harmonics can be calculated from:

$$Y_l^{-m}(\theta, \phi) = (-1)^m Y_l^{m*}(\theta, \phi) . \qquad (C.43)$$

Note that the spherical harmonics are normalised to unity on the unit sphere. They satisfy the orthonormality and closure relations and form a complete orthonormal set of square-integrable functions, as below:

$$\int_0^\pi \sin\theta\, d\theta \int_0^{2\pi} d\phi\, Y_j^{k*}(\theta, \phi) Y_l^m(\theta, \phi) = \delta_{jl}\delta_{km} , \qquad (C.44)$$

$$\sum_{l=0}^\infty \sum_{m=-l}^l Y_l^{m*}(\theta, \phi) Y_l^m(\theta', \phi') = \delta(\Omega - \Omega') . \qquad (C.45)$$

Index

Absorbed electrons 2, 3, 53
Angular momentum 5, 7, 13, 100

Backscattered electrons 1, 2, 56, 61, 69, 87
Backscattering coefficient 1, 53, 55–57, 61, 65, 67, 69, 76, 84, 87
Bessel functions 4, 27, 28, 36, 100, 101
Bethe formula 45, 51, 54, 75

Central potential 13, 17, 20
Continuous spectrum 97
Coulomb potential 20, 22
Cross-section 3, 17, 18, 20–22, 25, 33, 36–41, 47, 51, 54, 57, 65, 69, 70, 74, 76, 77, 81, 82

Density matrix 22–25
Depth distribution 1, 3, 60, 65, 82, 86
Dielectric function 45, 46, 76
Dirac equation 3, 10–15, 25, 26
Dirac notation 4, 6, 95, 98
Dirac–Hartree–Fock–Slater atomic potential 4

Eigenfunctions 91, 93, 94, 97, 100, 102
Eigenvalues 6, 7, 12, 15, 93, 97
Elastic mean free path 22, 74, 78
Elastic scattering 3, 15, 17, 18, 20–22, 25, 33, 36–41, 69, 74, 76, 77, 79, 81
Energy and angular distribution 56, 61, 69
Exchange effect 37, 47

Fermi energy 46, 48, 64, 81
First Born approximation 18, 20, 37, 74

Hartree–Fock atomic potential 3

Inelastic mean free path 4, 46, 47, 51, 76
Inelastic scattering 4, 79

Legendre polynomials 4, 28, 29, 99, 100

Matrix transformations 91
Monte Carlo 1–4, 65, 69, 70, 72, 73, 76–79, 81–86, 88–90
Multiple reflection method 61, 65–67, 84, 85

Operators 4–7, 9–14, 18, 19, 22, 23, 91, 93, 95–97, 102

Pauli matrices 8, 9, 11, 23, 32
Phase shifts 3, 28, 35, 36
Plasmons 2, 47, 48, 67, 69, 74, 87
Poisson distribution 72
Polarisation 22, 24, 33, 41
Positrons 1, 3, 15, 47, 65, 84, 86, 87, 89, 90
Pseudo-random-numbers 73

Random variables 3, 70–73
Relativistic partial-wave expansion method 3, 37–41, 76
Rutherford formula 21, 22, 37–40, 76

Schrödinger equation 10, 98, 100, 101
Secondary electrons 1, 2, 61–63, 69, 80, 81, 87–89
Simulations 1–4, 65, 72, 76, 77, 81, 82, 86
Solid state effect 37
Spherical harmonics 4, 30, 102, 103
Spin 3, 7, 8, 10–12, 15, 22–26, 31, 32
Stopping power 4, 44–47, 50, 51, 69, 70, 74, 76, 78, 81
Surface films 2, 66, 67, 77, 85

Thin films 1–4, 56–60, 65, 67, 69, 82, 84
Transmitted electrons 53, 69

Springer Tracts in Modern Physics

146 **Low-Energy Ion Irradiation of Solid Surfaces**
By H. Gnaser 1999. 93 figs. VIII, 293 pages

147 **Dispersion, Complex Analysis and Optical Spectroscopy**
By K.-E. Peiponen, E.M. Vartiainen, and T. Asakura 1999. 46 figs. VIII, 130 pages

148 **X-Ray Scattering from Soft-Matter Thin Films**
Materials Science and Basic Research
By M. Tolan 1999. 98 figs. IX, 197 pages

149 **High-Resolution X-Ray Scattering from Thin Films and Multilayers**
By V. Holý, U. Pietsch, and T. Baumbach 1999. 148 figs. XI, 256 pages

150 **QCD at HERA**
The Hadronic Final State in Deep Inelastic Scattering
By M. Kuhlen 1999. 99 figs. X, 172 pages

151 **Atomic Simulation of Electrooptic and Magnetooptic Oxide Materials**
By H. Donnerberg 1999. 45 figs. VIII, 205 pages

152 **Thermocapillary Convection in Models of Crystal Growth**
By H. Kuhlmann 1999. 101 figs. XVIII, 224 pages

153 **Neutral Kaons**
By R. Beluševi 1999. 67 figs. XII, 183 pages

154 **Applied RHEED**
Reflection High-Energy Electron Diffraction During Crystal Growth
By W. Braun 1999. 150 figs. IX, 222 pages

155 **High-Temperature-Superconductor Thin Films at Microwave Frequencies**
By M. Hein 1999. 134 figs. XIV, 395 pages

156 **Growth Processes and Surface Phase Equilibria in Molecular Beam Epitaxy**
By N.N. Ledentsov 1999. 17 figs. VIII, 84 pages

157 **Deposition of Diamond-Like Superhard Materials**
By W. Kulisch 1999. 60 figs. X, 191 pages

158 **Nonlinear Optics of Random Media**
Fractal Composites and Metal-Dielectric Films
By V.M. Shalaev 2000. 51 figs. XII, 158 pages

159 **Magnetic Dichroism in Core-Level Photoemission**
By K. Starke 2000. 64 figs. X, 136 pages

160 **Physics with Tau Leptons**
By A. Stahl 2000. 236 figs. VIII, 315 pages

161 **Semiclassical Theory of Mesoscopic Quantum Systems**
By K. Richter 2000. 50 figs. IX, 221 pages

162 **Electroweak Precision Tests at LEP**
By W. Hollik and G. Duckeck 2000. 60 figs. VIII, 161 pages

163 **Symmetries in Intermediate and High Energy Physics**
Ed. by A. Faessler, T.S. Kosmas, and G.K. Leontaris 2000. 96 figs. XVI, 316 pages

164 **Pattern Formation in Granular Materials**
By G.H. Ristow 2000. 83 figs. XIII, 161 pages

165 **Path Integral Quantization and Stochastic Quantization**
By M. Masujima 2000. 0 figs. XII, 282 pages

166 **Probing the Quantum Vacuum**
Pertubative Effective Action Approach in Quantum Electrodynamics and its Application
By W. Dittrich and H. Gies 2000. 16 figs. XI, 241 pages

167 **Photoelectric Properties and Applications of Low-Mobility Semiconductors**
By R. Könenkamp 2000. 57 figs. VIII, 100 pages

Springer Tracts in Modern Physics

168 **Deep Inelastic Positron-Proton Scattering in the High-Momentum-Transfer Regime of HERA**
By U.F. Katz 2000. 96 figs. VIII, 237 pages

169 **Semiconductor Cavity Quantum Electrodynamics**
By Y. Yamamoto, T. Tassone, H. Cao 2000. 67 figs. VIII, 154 pages

170 **d–d Excitations in Transition-Metal Oxides**
A Spin-Polarized Electron Energy-Loss Spectroscopy (SPEELS) Study
By B. Fromme 2001. 53 figs. XII, 143 pages

171 **High-T_c Superconductors for Magnet and Energy Technology**
By B. R. Lehndorff 2001. 139 figs. XII, 209 pages

172 **Dissipative Quantum Chaos and Decoherence**
By D. Braun 2001. 22 figs. XI, 132 pages

173 **Quantum Information**
An Introduction to Basic Theoretical Concepts and Experiments
By G. Alber, T. Beth, M. Horodecki, P. Horodecki, R. Horodecki, M. Rötteler, H. Weinfurter, R. Werner, and A. Zeilinger 2001. 60 figs. XI, 216 pages

174 **Superconductor/Semiconductor Junctions**
By Thomas Schäpers 2001. 91 figs. IX, 145 pages

175 **Ion-Induced Electron Emission from Crystalline Solids**
By Hiroshi Kudo 2002. 85 figs. IX, 161 pages

176 **Infrared Spectroscopy of Molecular Clusters**
An Introduction to Intermolecular Forces
By Martina Havenith 2002. 33 figs. VIII, 120 pages

177 **Applied Asymptotic Expansions in Momenta and Masses**
By Vladimir A. Smirnov 2002. 52 figs. IX, 263 pages

178 **Capillary Surfaces**
Shape – Stability – Dynamics, in Particular Under Weightlessnes
By Dieter Langbein 2002. 182 figs. XVIII, 364 pages

179 **Anomalous X-ray Scattering for Materials Characterization**
Atomic-Scale Structure Determination
By Yoshio Waseda 2002. 132 figs. XIV, 214 pages

180 **Coverings of Discrete Quasiperiodic Sets**
Theory and Applications to Quasicrystals
Edited by P. Kramer and Z. Papadopolos 2002. 128 figs., XIV, 274 pages

181 **Emulsion Science**
Basic Principles. An Overview
By J. Bibette, F. Leal-Calderon, V. Schmitt, and P. Poulin 2002. 50 figs., IX, 140 pages

182 **Transmission Electron Microscopy of Semiconductor Nanostructures**
An Analysis of Composition and Strain State
By A. Rosenauer 2003. 136 figs., XII, 238 pages

183 **Transverse Patterns in Nonlinear Optical Resonators**
By K. Staliūnas, V. J. Sánchez-Morcillo 2003. 132 figs., XII, 226 pages

184 **Statistical Physics and Economics**
Concepts, Tools and Applications
By M. Schulz 2003. 54 figs., XII, 244 pages

185 **Electronic Defect States in Alkali Halides**
Effects of Interaction with Molecular Ions
By V. Dierolf 2003. 80 figs., XII, 196 pages

186 **Electron-Beam Interactions with Solids**
Application of the Monte Carlo Method to Electron Scattering Problems
By M. Dapor 2003. 27 figs., X, 110 pages

Printing (Computer to Plate): Saladruck Berlin
Binding: Stürtz AG, Würzburg